The New Company Officer

THE NEW COMPANY OFFICER

Jason Hoevelmann

Fire Engineering
BOOKS & VIDEOS

Disclaimer

The recommendations, advice, descriptions, and methods in this book are presented solely for educational purposes. The author and publisher assume no liability whatsoever for any loss or damage that results from the use of any of the material in this book. Use of the material in this book is solely at the risk of the user.

Fire Engineering Books & Videos
110 S. Hartford Ave., Suite 200
Tulsa, Oklahoma 74120 USA

800.752.9764
+1.918.831.9421
info@fireengineeringbooks.com
www.FireEngineeringBooks.com

Publisher: Matthew Dresher
Managing Editor: Mark Haugh
Production Manager: Sheila Brock
Production Editor: Tony Quinn
Cover Designer: Charles Thomas

Library of Congress Cataloging-in-Publication Data

Names: Hoevelmann, Jason, author.
Title: The new company officer / Jason Hoevelmann.
Description: Tulsa, Oklahoma : PennWell, [2018] | Includes index.
Identifiers: LCCN 2017043245 | ISBN 9781593704117
Subjects: LCSH: Fire departments—Administration. | Fire extinction—Vocational guidance. | Fire fighters.
Classification: LCC TH9158 .H64 2018 | DDC 363.37068—dc23

Printed in the United States of America

4 5 24 23

Contents

Introduction

I never thought I would be one to write a book or offer anything substantial in writing. I never considered that anything that I had to offer would be used or shared by others. But, never be surprised by the community that is the fire service, for the brotherhood is strong and has ways of humbling you. With the encouragement and support of peers, mentors, and role models and the prodding of friends and cohorts, I take to penning (well, typing) my thoughts on why leadership is so critical to the success of the company officer.

Over the years I have had the good fortune to work for and with great leaders in the fire service and in my personal life. Not all of them were officers and most have never written a book, article, or blog post. They haven't taught in front of a full classroom at some big conference or contributed to a video series. No, most of them offered sound advice and then went about their business, never expecting anything in return. I guess you could say they weren't worried about accolades or "likes" on their Facebook page. A simple "thank you" and demonstrating and passing on that advice was good enough for them.

I would be very naive to believe that I have accomplished what I have at this point in my career without the many people who have influenced me. This book is a compilation of efforts, not even known to them, of fire service leaders who have crossed my path over the years. I am lucky to have had these individuals touch my life personally and professionally on my journey as a firefighter and fire officer. This book is as much theirs as it is mine and I couldn't be where I am today without them.

Some who have influenced me you have heard of, while other names you would never recognize unless you were from my small community. And yet, there are some people who I have no doubt had an influence on a decision I made or conversation that I had whose names I can't even remember, but their advice lives on in who I am because it was shared by somebody who cared enough about a young firefighter to say something or show me something.

My goal and intent of this book is for the reader to use and adapt what I have found to work well and to make it work for them. Not everything in here will work for everybody, but putting your own spin on a concept may produce just the right medicine to cure whatever problem you might be dealing with. You may disagree with some methods or thoughts, but at least put them to the test, challenge them to create success. Finally, share what you take out of this book with those who can use it. Write in it and take notes as you read. One sentence or phrase might jog your mind to something brilliant that *you* will come up with and use with great success.

I wish you all the best in your careers and thanks for reading. Remember, do the right things, the right way, for the right reasons, all the time.

1

PREPARATION

One of the questions I always get from firefighters wanting to promote and from students in the classes that I present is, "What can I do to get promoted?" That is not a simple question to answer and as many have said before, there are no silver bullets. There is a lot to consider when we look at that question and the conclusion is that there is no one thing to do to get promoted. There are many factors of varying importance that enter the equation of being promotable and getting promoted.

There are also variables outside of your control. Many organizations have different methods of promoting that can determine your fate without you having any impact at all on the results. There are many departments of all sizes and types that do not have legitimate promotional processes. Two examples of these illegitimate systems are politics and cronyism. Departments or districts with small governing bodies can manipulate and scheme to gain political favor. This happens often and can be very difficult to overcome without becoming part of the problem. Cronyism works the same way. Not being a part of a particular group, family, or demographic can exclude you from a legitimate opportunity of being promoted.

In other cases, it's just the process. In many of the volunteer departments in our country, promotions are still made through voting of the membership. This is a long-standing tradition and was meant to give control to the membership over who they wanted to lead them. It is a great idea, but human nature tends to take over and maneuvering takes place by those who are seeking positions they have no qualifications or experience for. The elections turn into a popularity contest.

This system also makes it difficult for officers who try to hold members accountable to remain in their positions: members who don't like being held to high standards or made accountable will either not vote them in at all, or create a movement and vote those officers out. This will cause the department to lose good, experienced, and well-meaning officers because they won't sacrifice their values and principles. Both the members and the department lose.

As we can see, the simple question of "How do I get myself promoted?" has no simple answer. Even in the best of organizations, getting promoted can be a stressful and long process. There are some who may never get promoted, but preparing for the position also prepares us for not obtaining it.

All is not lost—this chapter will address the things that will set you up for a successful run at promotion when the time is right and when you are ready. This discussion is *not* about creating an avenue for you to jump over others who are more senior, more experienced, and more qualified. This discussion is to offer you tips and methods to make sure you build a full body of work that is productive and conducive to being a strong leader and a great firefighter and member of your department.

Attitude

The new firefighter who walks into the engine house for the first time usually is a little soft-spoken and shy. They are very polite and eager to do whatever they're told. In some instances they are not sure whether to believe some of the things they're being told by the senior members or if they are being tested to see how gullible they are as the new kid in the house.

This is when your road to promotion begins. This is when you begin to become who you are and affect how others will think of you. This is where your character, morals, and integrity will come into play. And this is when you begin to determine what type of attitude you will bring with you every day to the engine house.

You notice I didn't say the kind of attitude created by having to do the grunt work for the first year or two as the most junior member. You notice I didn't say the kind of attitude running the multiple EMS calls for patients who "don't really need us" creates. You notice I didn't say what kind of attitude will be created by not getting any sleep at night when you have big plans the next day because you're out running the drunk and OBS calls. This is because *you* will determine your attitude, thus determining a part of *your* reputation. Only you can determine what your attitude will be.

Successful officers and leaders don't let circumstances determine their fate. Rather, they determine how they will use the circumstances to be successful. They will take the most difficult of circumstances and find a way to overcome. Resilience!

Not only will your attitude become a barometer that you are judged by, but I have found that eventually it will show up in how you deal with and approach the people you swore to serve when you took your oath. Many forget those words that were repeated with their right hand raised. For some it's just a formality during a swearing-in ceremony, but what it should be is a serious commitment that you recall frequently, especially when you are frustrated or thinking, "This isn't what I signed up for." Actually, yes it is! I encourage you to look back and reread what you swore to uphold and do. If you can't remember, ask for the words and keep them near or somewhere you can get to them easily. Any time your attitude begins to take a negative tone or you become egotistical or self-serving, pull those words out and read them again.

When you swore to serve and protect and uphold the policies, procedures, and guidelines of your department, there were no exceptions attached to that oath. Don't make exceptions for not providing your best and serving everyone you meet with compassion, empathy, and honor. They are yours to take care of and not judge, no matter how hard that can be at times.

"Don't make exceptions; be exceptional when adversity arrives."

Bad attitudes are easy to catch and can be hard to get rid of. Unfortunately, bad habits and negative attitudes are hard to change in the minds of those who promote and evaluate you. It becomes a part of your reputation (fig. 1–1).

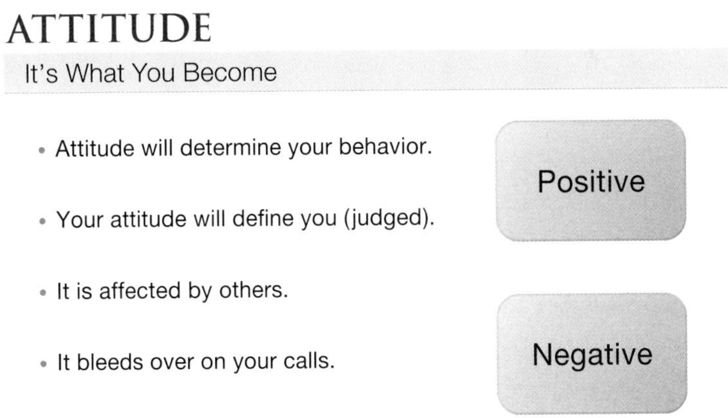

Fig. 1–1. Attitude—it's what you become.

"A good, solid, respected reputation takes a very long time to create, but only seconds or minutes to destroy."

When I say that this is where your road to promotion begins, I don't mean that you are living your life around the idea of being promoted; that's the wrong way to go about it and we'll discuss that later. Rather, I mean that you want to create a full and respectable body of work as a firefighter and as a person starting the first day on the job. What you are and how you act and react must be consistent. As you will hear again, you can't cram for promotion and you can't change your reputation overnight.

This section serves as the foundation for creating that consistent body of work that will be beneficial to you not only during times of promotional processes, but as a highly respected member and informal leader others look to for advice and guidance throughout your career.

You are going to go through times in your career that will challenge your ability and desire to do the right thing and to have that good attitude. Others around you are not going to share your mild temperament when things get rocky and they are going to encourage you to be like them. Don't let others who are miserable, disgruntled, lazy, mad, indifferent, or who have any other unproductive behaviors dictate your day. They don't deserve that kind of control over you and your behavior.

A final thought on your attitude: keep perspective. Our job is full of instances that are going to challenge your patience and ability to have empathy for the people you serve. They aren't always nice and in many cases they are not very appreciative of the service you provide. Some are going to take advantage of the system and they will use it for everything except what it is for. However, you didn't say you would protect and serve only the nice people or only the appreciative people. No, you swore to serve and protect, unconditionally.

We never know what others are going through. We don't know what struggles others are dealing with and they may use us as a comfort measure. It doesn't make it right, but our job is to solve problems, not to judge or to determine who is worthy of our services and who isn't.

In my career I can honestly say that I have seen some of the most severe punishments and discipline handed down because of three things: lying, stealing, and treating our customers poorly. Out of those three, treating our customers and citizens poorly is at the top of the list. Only bad things can happen when we aren't nice, patient, and empathetic.

Remember whom you're here for and whom you swore to serve and protect, and do it with genuine kindness and patience. In many cases, it's not necessarily what you say, but how you say it. We have all had that pointed out at some time in our life, whether at home or in our career. Saying the right thing with the wrong attitude, body language, or inflection is just as bad as saying the wrong thing.

Initiative

As little kids, initiative is not something we all have or aspire for. We would make a big mess with our toys and then leave it for somebody else (mom or dad) to clean up. Then when we were told to clean it up, we felt imposed upon. How dare our parents ask us to pick up the train wreck that is our bedroom! Remember those moments of awe your parents had when you offered to do some extra work around the house or to help your sibling without being asked? They would ask, "What do you want?" They knew something was up. But there were those kids who always made the rest of us look bad, who always did extra and were the exceptions to the rule. "Why can't you be more like Joe?" they would say. We hated that kid!

Fast-forward to adulthood. Some of us grew up and came to understand responsibility, and we understood that in the real world, things are competitive and very little was going to be just handed to us. We learned that hard work and doing extra could go a long way in separating us from the ones just doing the minimum. I realized this in the fire service at a young age. I wasn't always consistent until I matured a little, but I took notice of those who did extra and how that related to their successes and failures.

Initiative is important for you to build your reputation as a firefighter and future officer. It also generates respect from both your peers and your superiors, making you that informal leader who is so essential for any fire department.

Initiative comes in many forms and opportunities. It's up to you figure it out and listen when told. My wife works hard in whatever she does. She frequently overextends herself when she sees something that needs to be done. This is mostly when dealing with our church and school.

Our kids attend a small Catholic school in our town. The local Cub Scout pack never recruited in our school; they only went to the public schools in the area. When our oldest boy and his peers were old enough for scouts, they noticed that the local pack did not come to the private school like they did the public school to hand out information for those who might be interested in joining.

My wife gathered a few parents, one who had been in scouts as an adult, and worked hard, turned in piles of paperwork, recruited as many boys in our small school as she could, and was awarded a new pack through our parish. It turned out to be a great thing for not only the kids who attended our school, but the other kids of the parish who joined as well. It truly established a great family activity that no doubt benefited all of the boys who joined.

Additionally, when our school's playground equipment was old and almost unsafe, she created an avenue to buy new equipment at no expense to the school. She set up fundraising activities and applied for and received grants for the equipment, raising enough money to fully replace the old equipment with safer and larger equipment for our kids to play on. This took three years to complete! She not only took the initiative, she was resilient in completing a task nobody else wanted to take on.

Now, to say that she did this all by herself would be untrue. However, it took her getting the movement going for others to jump on and help. They saw that the efforts collectively could bring a positive outcome; they were willing to step up and offer their help. But it took one person's initiative to get others to join the movement.

She and others who worked on those projects could just as easily have sat around complaining about both issues and feeling sorry for our kids and school. They could have easily come to the conclusion that the problem was not theirs to solve or was beyond their reach or influence to change. But instead they took action—initiative! They solved the problem others either chose to ignore or did not see.

As a new firefighter or one who is looking to be attractive for promotion, initiative is an incredibly important trait. When I would listen to my wife's frustrations during these times and when she contemplates taking on other projects, I would ask her to let somebody else do it this time. Her saying is always, "See a need, fill a need." We must do the same thing in the firehouse.

When we arrive at the firehouse, whether it's for a career shift or a meeting or training night for the volunteer, once we put our gear on that apparatus, we own it. It is not the time to complain about who left the pack dirty, or who didn't change the battery in the radio, or any of the many things we see at shift change or when we arrive at the station. There is a way to handle those situations, but first you have to take ownership of the problem the community has and make it yours.

Each day or shift or activity we spend at the firehouse must be looked at as an opportunity to learn something. We need to take the initiative to ask productive and relevant questions with the pure intent of learning something new. Ask "Why?" when you don't know. There is a great deal of experience in every firehouse and hall in the nation that we must tap and absorb. They have stories from calls that will explain something you don't fully understand. The key to learning successfully, however, is being respectful and genuine with your questions and following and applying the advice and answers given.

To quote Banksy, "A lot of people never use their initiative because no one told them to." Your motivation for taking action and showing initiative should come from the expectation that doing the right thing is the right thing to do. The motivation should never be external. What I mean by that is that it shouldn't matter who notices or gives you a pat on the back. If it's the right thing to do, that should be enough. Doing the small things will be noticed by those around you and by your officer over time.

Nobody, especially an officer, likes a firefighter who is continually letting everyone know what great things they've done. The other crewmembers will get tired of it and the officer will get tired of it. It shows that you are more concerned with *you* than you are with making sure things get done the way they need to get done. Remember, you are not the center of the universe, only an active participant. But you can have a positive impact by exceeding expectations without exceeding the patience of others with your "look what I did" attitude.

Do things because they are the right things to do, not because you might get noticed. Do the things that nobody else wants to do and do them excellently! See a need, fill a need (fig. 1–2).

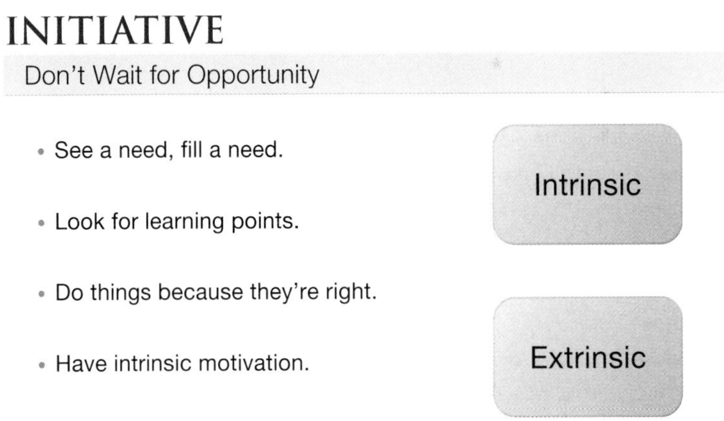

Fig. 1–2. Don't wait for opportunity.

Learning

We must take every opportunity to learn. This is not necessarily going to classes, attending conferences, drilling, or training, although all of those things are important. Learning is as much about paying attention as it is about taking action through the items I just listed.

Every fire department has its senior members, middle members, and junior members. Each one has seen something you haven't. Each one has performed a task or technique that you haven't. Each one has had successes and failures that you have yet to experience. These are critical for us to learn from.

We do this in two basic ways. One way is just to listen to stories and discussions that these other members offer during training evolutions and drills or just when sitting on the front bumper or at the kitchen table. The other way is to ask! As I mentioned while discussing initiative, you must be respectful and genuine with your questions and listen intently when the answers are provided. Someday all of that experience and knowledge will walk out the door, making it more difficult to tap that source of valuable information.

Once you ask and listen, you must use the information. It is useless for you to ask an officer or senior member for advice that is sound and tested and then ignore it because you think you know better. There is no quicker way to become irrelevant to those individuals than to ask for help or advice and then completely ignore it. Be respectful and put that experience to work.

A big mistake I see firefighters make, especially when working toward promotion, is that they don't spend enough time on mastering the basics of firefighting. There cannot be a large enough emphasis placed on mastering the basics that are used on every fire or incident. These skills are often overlooked and put on a back rack in favor of more advanced or technical skills, many which are seldom utilized.

I had a firefighter come into my office one day who had recently become eligible for promotion to the rank of captain. There wasn't a current opening available and there wasn't going to be one anytime soon. He came in and wanted to know where he needed to work to make him a more attractive candidate for promotion. Of course, many things came to mind, but for him specifically, my answer was easy, one given to me by mentors and one I have told many others. I asked if he was willing to commit a great deal of his own time to become a better firefighter. Of course he agreed that he was. The simple answer I provided was to get as intimate with every part of being a firefighter as he could. Not really the answer he was looking for and not an answer that is easy or that can be done quickly.

He gave me a little bit of puzzled look. I continued: "When anything goes wrong with your apparatus, who will your crew likely go to first? The officer. When a piece of equipment doesn't operate like it should, who will they go to? The officer." I gave a few more examples and asked if he was currently prepared to troubleshoot any problem confidently and quickly right now? He thought about it and said, "No."

Not that you will ever know everything about everything, but you should be close when it comes to your equipment. You should know exactly how much each line should be pumped at and what the capabilities of your nozzles and hose are. You should know how to force a door with irons without having to think about it. Can you do it in a tight hallway or in the dark? Do you know how long you'll need to stretch hose at the garden apartments a few blocks down the road? Do you think about backup plans in the event that the initial actions aren't successful?

As a firefighter seeking promotion or even as a company officer seeking to advance, be the best at what you're doing *now*. Be the best firefighter in the department, don't try to be the best company-officer-to-be. Be the best company officer, not the best chief-officer-to-be. Master the job you currently possess. I've seen too many firefighters and officers worry so much about the next rank or position that they lose their perspective and struggle doing their current job well. Instead of excelling at their current position, they struggle and end up falling behind somebody who stayed consistent and became the best at what they do every day.

Another trap of thinking ahead rather than keeping your mind on your present position is spending large sums of time and money on training that will benefit neither you nor the department. There is nothing wrong with broadening your horizons regarding training and education, and I'm not saying you shouldn't take classes that don't directly relate to your job description. What is important is that you must understand the needs of your department and community. If your department does not have high-angle hazards, you probably should not invest greatly in high angle rescue training. Firefighters and officers will dive into a specialty with great enthusiasm and vigor even though their department cannot offer the resources or equipment to actually perform in that capacity because they don't have the hazard in their response area, lack funds, lack additional personnel to effectively perform in that capacity, or all of the above.

Unfortunately, that enthusiasm and vigor are not always given to training for the types of incidents that are actually responded to. Some of these firefighters feel that training to force doors with a Halligan bar and a flat-headed axe is below them because they did it last year. It is inexcusable if there is an inability to effectively pull a line to a second-story apartment or an inability to stretch a static line into an apartment building.

The point is this: train in a way and on topics that are going to benefit you, your crew, and your department, not just on specialized skills that provide you with a feel-good diploma on your wall and padding on your résumé. Have a purpose and make sure you are doing what's best for the department as well as yourself. It's the responsible thing to do and you must remember where your loyalties are: with your department and its members.

When I took over as the training officer in our department (which previously had not had any real training division), I had a flood of requests for all kinds of training. I had secured a respectable budget and full support and autonomy from the chief and our board of directors and it felt like open season for those who wanted to take a variety of classes.

I immediately had a few members in my office who wanted to get nursing degrees. We are a typical EMS-based fire district that only responds to 911 calls and does not do transfers. Being responsible for the budget and understanding how the fire service works, I knew that if I just blindly approved these requests in the name of training and education, I would end up with more requests that pushed the envelope in regard to what related to our response capabilities and resources. So I asked each one of them how obtaining a nursing degree would benefit the department, the members, and our citizens directly. They tried to tell me about how they would have more knowledge about prescription medications and a few other items that didn't hold a lot of water. Nothing they could tell me added value directly to how we operate. The only benefit I could find was that they could get a real good part-time job on their days off. I denied the requests and over the years have had to deny others using the same formula.

Along with identifying what your department is doing now, look at where it might be going when trying to identify training needs. Is EMS in the future? Is a new industrial facility being developed that might require a hazardous materials response? If you keep tabs on what is happening in your department and community, you may find that there are other circumstances that provide an opportunity to become a huge asset.

You're going to go to conferences and take classes from instructors who work in very different departments than yours. If you belong to a smaller, suburban department or a small, rural volunteer department, you must keep that perspective in mind. One thing to consider is that your department was doing things for a long time before you arrived. That doesn't mean they've been doing it all correctly, but it doesn't mean they have been doing it wrong either. Be patient coming back from these training opportunities. Present new ideas in a respectful and productive manner.

Don't completely discount your interests and goals, though. If you are seeking promotional advancement, you will need to attend and participate in officer-level training and courses. If you are consistent and you plan, you can accomplish this with relative ease while still mastering your current position. It's important that you spend some time doing what you like. You can take free classes in subjects that interest you via resources such as EdX on the Internet. Study topics to become a more diverse and well-rounded member of your fire department. Take some time for you and take some classes for you, but just don't forget the mission.

One really important factor regarding your participation and planning for taking officer-level classes is that you should expect to surpass the minimum requirements that are currently in place for officers. The minimum requirements for promotion to captain often look something like this:

- Fire Officer 1
- Associate's degree
- Fire Instructor 1
- Fire department safety officer
- Incident command system (ICS) 100, 200, 300, 700, and 800

Unless you are going to be promoted immediately, you should exceed those requirements. It is not uncommon for departments to raise the minimum standards to increase professionalism and raise the bar over time. This is done for the department's own sake or sometimes done to reduce the number of candidates. Your goal may look something like this:

- Fire Officer 1
- Fire Officer 2
- Associate's degree
- Fire Instructor 1
- Fire department safety officer
- ICS 100, 200, 300, 400, 700, and 800
- Command and control of emergency incidents

This will give you a head start if the requirements do change, and it is prudent to expect them to. If you're not sure what additional classes you should be looking at, ask your chief or look at the next rank above you, the one you're pursuing, and start attending those classes. Some people will complain about the requirements changing—don't let yourself fall into that trap. As with any profession, expectations change and should be raised to meet the challenges of our service. Stay a step ahead and be consistent and the changes will be easy for you to deal with and accept (fig. 1–3).

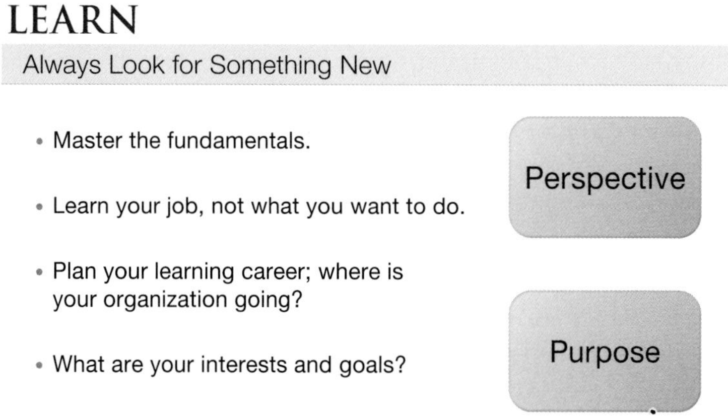

Fig. 1–3. Always look for something new.

Responsibility

There is always a lot of talk in officer-level classes and leadership classes about responsibility. We throw the word around in discussions at the kitchen table, and we have listened to many officer candidates during promotional interviews offer responsibility as something they would ensure was taken if they were promoted.

The truth is that we all don't take enough responsibility when it matters. It's very easy to be a glory hound and take responsibility for all of the good outcomes, whether you had anything to do with it or not. This is a fast way to lose your crew and members. Remember, you serve your members, so when things go well, they should get the credit and when things go poorly, you need to publicly take responsibility. You can deal with individual failures in private afterward.

It's also easy to take responsibility for something when the consequences are light. When the outcomes, even if negative, are not severe, taking responsibility is very easy and without risk. In fact, you will see individuals try to capitalize on taking responsibility in these instances just to make themselves look good. It's just as bad as only taking responsibility for the good things.

A true leader earns their stripes by taking responsibility when things are bad or difficult, even complete failures. These are the times when we see fewer people sticking their necks out. What truly separates those who take responsibility because they know how important it is and those who want accolades are the situations that each accept as "theirs."

Responsibility comes in many shapes and forms and situations. One area that people fail in is owning the situation. Whatever mistake or failure has occurred, many will try to blame others or just ignore the problem. They are more concerned with looking bad or being considered inferior, so they refuse to admit to themselves that they have made a mistake and/or need to be the one to fix the problem. When it is the officer who fails to take responsibility or to take ownership of problems or failures, that mentality can create a subculture in your department and become systemic. This is a *huge* problem. If firefighters are unwilling to admit failures or mistakes and take responsibility for them, we begin to cultivate a culture of "me" instead of "us." It really is critical that as future and current officers, and just as decent people, we own our problems and admit our mistakes when they happen. We have to create an atmosphere and environment that encourages not only admitting and owning our own mistakes and failures but also owning whatever we find that needs to be corrected or handled.

There is no skirting responsibility in our profession. We can't afford it in any way, shape, or form. There is too much at stake to not have the integrity to admit when you make a mistake or when your crew or members make a mistake. Taking responsibility is really about being honest with yourself and taking action when it is not pleasant or easy (fig. 1–4). An officer who takes responsibility will eventually have a crew that takes responsibility because it will be expected.

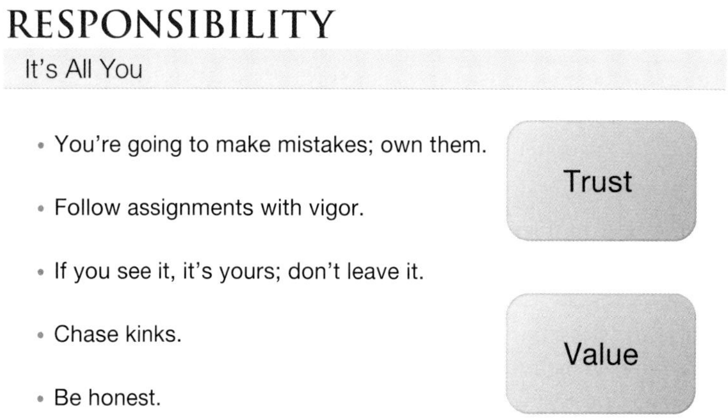

Fig. 1–4. It's all you.

Set the Example

This concept and theory is probably one of the most misunderstood and ill-defined of the sections in the chapter. Not because of effort or attention, but because of the generalities that are frequently attached to the concept of "being an example."

One of the first comments made when a firefighter is asked what kind of leader they would be is that they would lead by example. It's a common response in the leadership class I teach, during promotion interviews, and during coaching and mentoring sessions I have with firefighters and officers looking to advance and improve.

It is commendable to list leading by example as a way you would lead. However, it must be applied appropriately and along with other traits that leaders must possess to be successful. In that, you must have specific goals and objectives in mind on *how* you would lead by example. It is not enough to simply state this; leading by example is more complicated than dropping a buzzword or having good intentions.

When I do ask members and potential candidates to expand on how they would lead by example, the typical response involves cleaning and doing chores with your members. That is not wrong—in fact, it is a good idea—but it will only get you so far. That task, more than likely, is something your members will expect you do anyway.

In this section we are not really talking about *leading* by example as much as *setting* the example. There is a difference, and I believe you must set the example before you can lead by it. It is a progressive process that can't be ignored. And if you set the example on a consistent basis throughout your career, leading by the same will be more natural and you will have a better idea about what that means and understand the benefits and challenges of doing so.

Setting an example can be good or bad, and in the fire service, at times, we condone the bad examples over the good because of the culture of our organization. A perfect example of this is helmets. Our traditions give a lot of respect and validation to members whose helmets are blackened, crispy, or otherwise showing signs of being in some hot, smoky fires. I'll be the first to admit that I like the look of a battle-worn helmet—it does say something about where that helmet has been and what that firefighter has seen and experienced. Firefighters are very attached to and proud of their helmets, as they should be. However, they can take it a little too far, particularly now that we know that contact with the contaminants stored on dirty helmets can cause cancers. It is not an opinion, it is a fact of today's fire service. But firefighters still refuse to clean their helmets.

Setting the example then, in this instance, would be being the first to clean your helmet and gear after a fire. You're making it "cool" to clean your stuff after a fire. It not only demonstrates an awareness of the big picture, but I also used to tell others who gave me funny looks when I was washing my helmet and gear, "Eh, it's no big deal, we're going to get the next one too, so it'll get darkened again." Setting the example in this situation will bring you to leading by example. There are many ways a firefighter sets the example for others that can set the firefighter up for being that informal leader before they are ever promoted.

The question gets asked, "What can a person do to be first on the promotion exam and process?" Askers are looking for advice on how to interview, what to study, and how to practice on scenarios for the simulations and tabletops. You have to perform well on those aspects of the process, no doubt about it. But your reputation and the example you have set will play a large role in your success or failure in the promotion process.

When having this conversation with candidate officers I like to give them a list of other "firsts" to worry about before they ever take the test. It is a list I have created and added to over the years as I observed my own behaviors and those of others who were working toward promotion. The following list is in no way complete, nor is it perfect. There are obvious hurdles and challenges for these based on who you work for and the culture of the organization. But by and large, being first in these things will greatly increase your odds of building a consistent and strong body of work you can refer to when promotion time comes around.

The List of Firsts

- Be the first to report for work every day.
- Be the first to check, clean, and inspect equipment.
- Be the first to pick up and start chores.
- Be the first to offer to do unattractive tasks.
- Be the first to demonstrate or start a drill or training evolution.
- Be the first to offer to help with training and drills.
- Be the first to offer help to other members in any and all tasks and functions.
- Be the first to the apparatus on *every* call, not just fires.
- Be the first to speak up when something is wrong, but respectfully.
- Be the first to ask questions that are genuine and inquisitive, not confrontational.
- Be the first to do the right things, the right way, all of the time!

This list sounds simple and you might say that every item on that list is common sense and that most people know this. But it's interesting that so many don't follow the advice of this list. In fact, some will do everything in their power to avoid being first in those items.

It's like working out or needing to get a little healthier. For example, a guy has high cholesterol and has for several years. This person never believes that he will have any adverse effects from the condition and continues to eat what he wants and doesn't exercise as frequently as he should. Life is good until one day his blocked arteries give him a heart attack. Now all of a sudden eating better and performing a little exercise is important, and he wishes he had done things differently for the past several years. But not until he had an emergency did he believe he needed small, consistent changes and compromises in his life: dieting and exercise may have been uncomfortable but would have saved him in the future.

I have had firefighters come into my office or meet with me asking for promotion advice and when I propose this list, they nod their heads but never fully put it in practice. Then one day a promotion comes around and they get passed over for the person who did follow the list of firsts. It's like the heart attack: they were told it was coming and how to avoid it, but they didn't think it would happen. They weren't willing to do extra and to be a little uncomfortable in order to avoid the consequences.

It's common to see a mad dash to engagement and going the extra mile when a promotion is announced or is expected a few months down the road because of a retirement. Things get real interesting in the fire service and people you thought you knew well start doing crazy things. Those who normally prefer the recliner are now up and active all day just looking for something productive to do. Members who prefer their bowling league and other off-duty activities are now registering for class off duty with no pay, just to learn! The member who would never read a book and has accused you of being a "fire nerd" because you routinely read the trade magazines and books is now coming to you asking you if they can borrow some your periodicals. Does this sound familiar?

These members are cramming for a promotion, and it works about as well as cramming for a college exam when you've played hooky. They have not done the consistent work required to set an example and be considered for the upcoming promotion. These members almost always revert back to their old ways in time when they don't get promoted. What amazes me is that they're surprised!

Don't cram for promotions! Be steady and consistent with your actions and intentions. Setting the example comes in limitless forms for those who really wish to do this job and engage themselves in our profession. Some ways are small and hard to notice without looking for them. Other ways are significant contributions that are seen by everybody. Whether large or small, the actions and intentions of your job should be to serve. By doing that, you will set the right kind of example for everyone who comes into contact with you, no matter rank or years of service.

2

FIVE COMMON MISTAKES NEW OFFICERS MAKE

This topic is appropriate for the simple fact that I get asked at a lot of classes and conferences, "What are the most common mistakes that new officers make?" Secondly, I like to learn from other's mistakes and successes, so why not include some observations and struggles that officers have made and had? We should absorb all the knowledge we can to do our jobs better (fig. 2–1). This list is not all-inclusive nor is it absolute. You may have your own list with unmentioned mistakes. Some of these issues are organizationally driven and need to be addressed beyond individual choice. That's perfectly fine. If you do have other issues that belong on this list, share them and encourage others to learn from them. Don't hoard that information all for yourself.

Don't Try to Change Everything

The first mistake we see new officers (and some current officers) make is trying to change the world they work in as fast as possible. You've got all these ideas and you've taken all these classes and you've done all this training and you've read these books and articles and blogs—it's easy to come into a situation and try too hard to push the kind of change you believe needs to happen.

Sometimes you have to go with the pace of the organization or just barely push ahead. You can't always be way out ahead because you're going leave people behind and they're going to get frustrated. Once that frustration sets in, they're going to be very unhappy and they're not going to be responsive to the quality changes that you're trying to implement.

Fig. 2–1. With the red helmet comes new responsibilities.

It can be very difficult to make sweeping changes all at once. It's more important to come in and make small but significant changes in a timely fashion. You must be aware of what your organization is about, what kind of situation it's in, and push those changes as appropriate. You're going to have to be adaptable and flexible, and your ideas may have to be reined in just a little bit in order to meet the needs of the group and your organization. It can't be just an all-out "here are all these things we're going to do differently" and damn the consequences.

Something else to consider is that it's not always a good idea to come in and start changing everything when you have replaced an officer, especially if your crew liked the former officer. What you're essentially saying is that everything that crew and that officer have been doing is wrong. It can create an immediate barrier between you and your new crew.

Currently, and for several years now, the big catchphrase has been "We're going to change the culture," of a department specifically and the fire service in general. It's easy to say, but it's much harder to do. Use caution when attempting to come in and change the culture of a shift or a crew, and definitely when you come to change the culture of a department.

If you want to change culture, you're going to have to do it in very small increments and start with changing individual behaviors. Don't rush in trying to change the culture of any one of those environments; first look at what individual behaviors need to be changed and change those one at a time, which in effect will change the culture. But it takes a long time to do. Changing anything in the fire service takes a high degree of patience and steady progress.

Don't Think Too Much of Yourself

Don't come in with an authority complex saying, "Hey, look at me, I'm a new officer and you have to do what I say," "Hey, look what I did," "You know, it's about time I got promoted," or any other type of self-indulgent attitude. You should come in like a lamb, not so much like a lion. If you have to push your authority onto others, you probably shouldn't have been promoted and your advancement was a mistake.

Take it for what it's worth, but the best officers and leaders come in and they pick up where the others left off. They understand the job description. They understand the objectives. They create goals and objectives for their crew and their shift and they discuss those things. They don't have to come in, flash the badge, flash the new color helmet or other rank-change signifiers just to prove their worth and authority. Those things don't have to be expressed other than in the fact that you're doing your job and you're setting expectations for your crew and your shift. As you promote up, the importance of your position is not based on how many people work *for* you, but rather how many people you are getting the privilege to serve. Many get it backward and think that because they've been promoted, because now they have rank and title and formal authority, that everybody should be taking care of them. In reality, instead of other people serving you or working for you, the higher up you promote, the more people you have to serve and the more people *you* work for.

Putting others before yourself and taking your promotion quietly without an increase in your ego will go a long way in building respect and getting buy-in for the changes and expectations you set for your crews.

Don't Micromanage

Micromanaging kills! Nothing will ruin your ability to have relationships and to build trust and credibility as a fire officer more than micromanaging your crew. A key component for any officer, any leader, is the ability to delegate appropriately and to offer responsibility to those who can handle it. It doesn't mean they're going to do it the way you do it or the way you did it or the way you think it should be done. The bottom line is if we're going to be effective leaders and officers, we have to delegate effectively.

Delegating is good for morale. It's also good for up-and-coming leaders and future officers to be coached in tasks. It's also good for officers because it passes on tasks so officers can accept new responsibilities that only they can deal with.

It can be hard to pass jobs and functions you're really passionate about on to others, but we have to allow ourselves to stand back and let others learn, succeed, and fail. A key component of delegation and not micromanaging is understanding that sometimes our crew members are going to fail and we have to allow them to do so when it's appropriate.

A huge component of being an officer and an effective leader is building relationships and trust with your crew members. Part of that relationship is that you trust them to do the job—after providing resources, training, knowledge, education, and mentoring. You trust those crewmembers because they're an extension of yourself.

Don't Quit Working

"Don't quit: it's not the finish line!" A chief once told me that; it was great advice. What that means is that when young firefighters become eligible for promotion, at whatever rank that might be, they really apply themselves and get engaged. They read, take classes, and train, they take on responsibilities at our fire departments, they take on extra tasks, ask for advice, and they do a lot of things, hopefully over a long period of time, to build a consistent body of work that's conducive to them being an attractive and quality candidate for promotion. Along with that comes a lot of energy and a lot of work.

There can be personal sacrifice when they take classes and spend weekends going to conferences. They read articles instead of the novels that they like. They put a lot of time and effort into making themselves better all around. Then, for whatever the reason, they get promoted and slow down; they quit. They think they've achieved what their end goal was, but the end goal is never the promotion.

If we quit after the promotion, we're not going to have the knowledge and mastery we need to have. We're not going to have the problem-solving abilities that are so critical for us to have. We're not going to know where to find the resources when we don't have the answers. We're going to become stagnant and complacent and it will filter down to our crew. Many newly promoted fire officers think they've crossed the finish line. They broke the tape in the race and now they get to sit back and cruise. But in reality, you should be working even harder at getting better and staying better as an officer than you did as a firefighter.

Even if you mastered all of the things you need to master as a firefighter—which is highly unlikely, but maybe in your mind, you've put every ounce that you could into being the best firefighter you can be—now you've got to do the same thing as a company officer. This means still maintaining those basic fundamental skills and adding all of the tasks and knowledge you need to be the best fire officer you can be. This means attending classes, reading books and articles, going to conferences, and asking for advice—all of the things you did as a firefighter trying to get promoted, you need to do as a fire officer to be the person others can come to and ask for that same advice and information. When you get promoted, although you've achieved much, now it's time to set new goals.

Don't Avoid Discipline

A huge pitfall to being an effective officer is discipline. Most of us have known the people we work with for a long time. Buddies can be the toughest to deal with sometimes; it's never easy to hand out discipline to friends and close acquaintances. You'll experience this the first time you tell somebody no and you get a quizzical look.

When you do stick to your guns and hand out discipline when it's appropriate, they may accuse you of forgetting where you came from. Don't second-guess your actions. You simply have different responsibilities that include keeping members inside the established boundaries of the department.

With these new responsibilities and authority, you have to be accountable. When people screw up you now are the one who has to hold them accountable, and if you don't, you're going to become a pushover. Yeah, they may like you. They may think you're really cool to hang out with, but you're not going to have the respect or the admiration of those who understand how a shift really works. You have to be able to tell people no and you have to be able to hand out discipline.

It's fine to have a conversation and most of the time that works. Once you set your expectations and you lay out what is expected of your officers, it's pretty easy to fix and change those behaviors. But some people need a little extra motivation to not do the same things over and over again; it may be something they've gotten away with for years because past officers wouldn't address it. Don't buy into the whole "Well, that's been precedent." That's a bunch of BS. If it's wrong, it's wrong. If it's right, it's right, and as the officer you have to make sure you correct and mitigate it.

Discipline, especially to friends, is going to be one of the hardest and most unpleasant things you ever do in your career as a fire officer. Could it change your relationships? It might. Some will say to not take it personally. You may take it personally anyway and so might they, but the important thing is that you can't act on those personal emotions. You have to act on the expectations from your superior officers, what the missions and objections and goals are of the organization and, most importantly, what's right.

What is fair to everybody? What's fair to the rest of the crew? What's fair to the people who do it right every single day? If you let people slide on a regular basis, eventually everybody is going to want that same consideration when they screw up and then you've got a lack of control and discipline and it makes it that much harder to fix it. As soon as you get promoted you have to make it clear: "Listen, I know people have gotten away with this in the past, but we're not doing that any longer."

Don't be afraid of and don't avoid discipline, especially with people who are friends. That's the quickest way to get yourself into a great deal of trouble. You don't have to be quick to hand down hard discipline. When it's appropriate use it and when it's not, don't. Just don't avoid it and don't let things slide because it's a real slippery slope.

3

SETTING EXPECTATIONS AND PRIORITIES

As a new officer in my volunteer fire department, I thought being an officer was just giving orders and being "in charge." I quickly learned that I was mistaken. Without clear direction and boundaries, members are just guessing what you want and what they are supposed do. We can tell them what to do every step, but that is also not effective or efficient. I learned by experience, by taking classes, and by listening to some smart fire service veterans that to be successful as an officer I should establish clear expectations for my members.

Setting expectations is like creating a road map for how members are to act, operate, and serve (fig. 3–1). It is not a way to micromanage; there is a distinct difference between providing direction and being the one who has to always be the driver of the car.

Expectations should not be something you come up with at the last minute. They are things you should be thinking of as you start to consider participating in the promotional process. The old saying—I had an officer tell me this—is, "Just do your job and don't get me in trouble," but that's not good enough nor is it fair to your members. That is not an expectation, that is a lazy officer who has no interest in taking responsibility for their members and does not have the mission of the organization in mind.

Being a new officer is an opportunity to set the tone and create the boundaries in which your members are to operate. Having a meeting is a chance to put into place a new attitude, atmosphere, and behaviors as needed as a new officer. However, this is not a task for only new officers; it is also highly recommended that veteran officers have this meeting to set expectations if they haven't already done so.

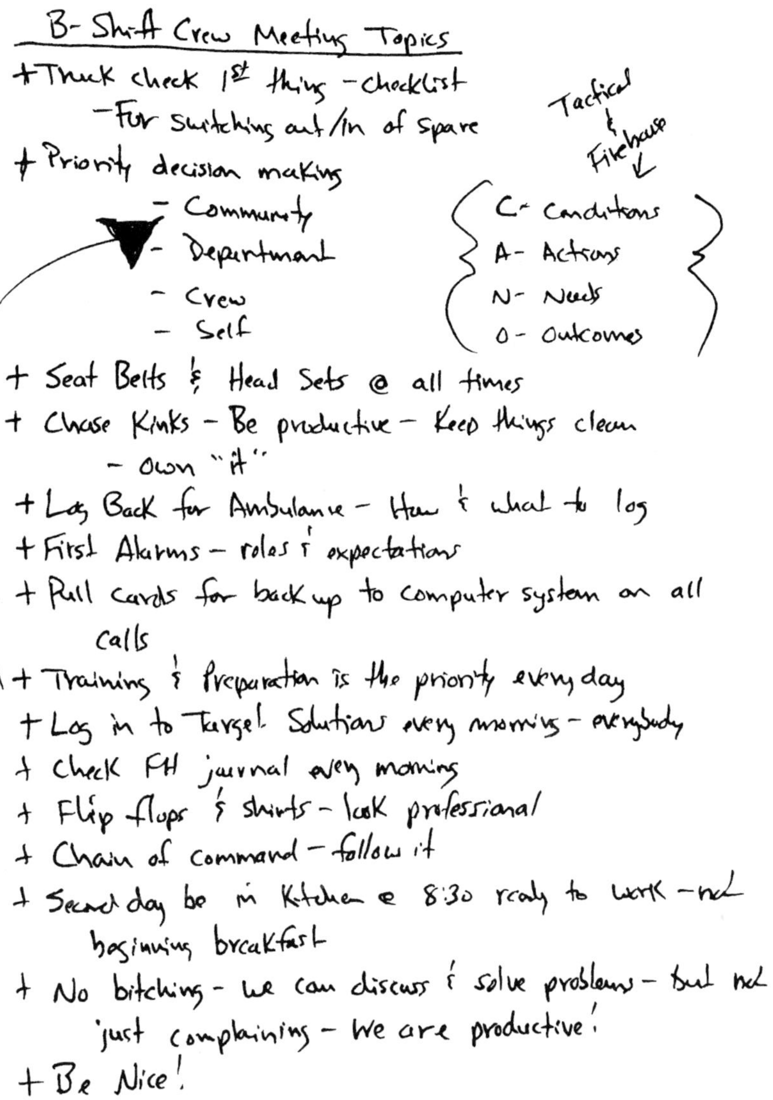

B-Shift Crew Meeting Topics

+ Truck check 1st thing – checklist
  – For switching out/in of spare
+ Priority decision making
  – Community
  – Department
  – Crew
  – Self

Tactical & Firehouse

C – Conditions
A – Actions
N – Needs
O – Outcomes

+ Seat Belts & Head Sets @ all times
+ Chase Kinks – Be productive – Keep things clean
  – own "it"
+ Log Back for Ambulance – How & what to log
+ First Alarms – roles & expectations
+ Pull cards for back up to computer system on all calls
+ Training & Preparation is the priority every day
+ Log in to Target Solutions every morning – everybody
+ Check FH journal every morning
+ Flip flops & shirts – look professional
+ Chain of command – follow it
+ Second day be in Kitchen @ 8:30 ready to work – not beginning breakfast
+ No bitching – we can discuss & solve problems – but not just complaining – We are productive!
+ Be Nice!

Fig. 3–1. It is crucial that you set expectations with your crew.

"When should I have a meeting to set expectations?" is a question that is frequently asked in classes and during conferences. There are a few recommendations for how to do this. The advice being offered here is in no way the only way to do this, just what I have found to be effective.

It's best to meet fairly early with your new crew. Don't let much time go by without at least covering some of the most important expectations you have created. The problem with waiting is that you will see things you don't like happening and say to yourself, "I'll wait until we have our meeting to address this," which can cause some problems.

There is some validity to waiting to "feel out" your crew and figure out what needs to be addressed. There are some things you can choose to wait on, but generally speaking, you should have some prior notice of your promotion and your expectations should be based on what you expect, not necessarily what you observe or witness. Although you could argue that you can better identify problems if you wait, noticing and addressing problems is going to be a continuous effort, while you can't wait too long to start sharing your expectations.

Having had the opportunity to try both methods, meeting with the crew immediately proved to be the right decision. It set the tone right out of the gate and created those boundaries that are needed immediately. There was no question from the first day about what my expectations were and where we stood as a crew.

You're going to want to write up a list of your expectations prior to getting your crew, then meet with your crew as early as possible. This eliminates a number of questions, assumptions, and speculations. Let's face it, they have a preconceived opinion of you and you may have the same of some of them. The earlier you can address and meet with your members, the sooner both assumptions can be dispelled (or confirmed).

So where do you start? The first thing you need to do is have a plan. This cannot be a shoot-from-the-hip situation. As was mentioned before, you should have some idea about what your fundamental beliefs and core values are. Put some thought into this and create a written list of expectations to share with your crew. Some will be reinforcements of items your crew already knows or of department policy, while others will be unique to your crew and your ideals. Your expectations should line up with your department's mission as well. Don't go all out in left field with your own ideas and philosophies if they are not in line with the department's own mission; it will only cause confusion for your crew and very likely some grief for you.

After you get your list figured out, build justifications for each item on your list. Be able to defend and specifically define each expectation you give your crew. You will be challenged on a few and you will need to have a specific reason and an example to defend it. In the large majority of cases, "because I said so" isn't enough to get buy-in and understanding. It is critical that you be able to provide the why for every item on your list; if you can't defend it, it probably doesn't belong on the list.

When you nail down your expectations and are ready for the meeting, let your crew know ahead of time. Set a time and date. If in a volunteer house, it's no different: set a time and day and make sure it's just your crew, house, or station personnel. If others are there, wait for them to leave or ask them to leave—respectfully, of course.

As the captain, let your battalion chief, chief, or shift chief know what you're doing and that you are requesting privacy. Having this discussion with your battalion chief prior to the meeting and going over the list of expectations ensures that you are not overstepping any boundaries of theirs or the department's. It's an essential sign of respect to keep them in the loop and ensure that there are no contradictions to any expectations or ideas they might already have. This is a critical step in creating credibility and respect with your superior officer. Don't skip this step, even if you have problems with this person. It will only make things worse if you go around them.

You can't have distractions, other than running calls, so limit them as best you can. It is recommended that you *not* do this offsite and off duty. This needs to be a department event. When this meeting takes place offsite and is not during duty or business hours, it can lead to some skeptics causing problems during the meeting or members leaving or not showing up at all. Hold your meeting on department time to keep it official departmental business. If you are a career officer dealing with your crew, it is recommended that you wait until the offgoing crew is gone. We know how the morning shift change can have a few hangarounds drinking coffee and gossiping. In truth, it's normally a good time. However, they don't need to know all of our business, so wait until they leave to conduct your meeting.

Plan to address the topics that will have the most discussion or discord toward the end of the meeting. There are two schools of thought on this, and those who argue for addressing the thorny issues first say that you want to make sure you get them covered before members get calls or in case the meeting gets too long and you lose their attention. Others, myself included, like the idea of waiting until the end to address these items so we can quickly get through the easy ones, leaving ample time for productive debate and discussion for the more difficult or controversial items. When there are more topics to cover, there is a tendency to rush through the tough topics to ensure that the entire list is addressed. I have had success with covering the easy stuff first and keeping things moving on the front end. It can put some members at ease and gives them a chance to get comfortable by asking about easy items before the tougher subjects come around.

Finally, let it rip. Be concise and honest and stay on point. Go in and use your list. That's not to say you can't discuss something if it is important, but you need to stick with what your original intentions were. It needs to be noted and stated that while some items can be discussed and there may be room for compromise, some of the items are not open for changes or variances. You must create those boundaries early on.

If there are expectations, there need to be consequences for not meeting them. This needs to be addressed early in the conversation as well. Consequences don't necessarily mean discipline. If we don't perform well on a call or in a drill, the consequence may be that we rearrange our schedule to accommodate more drilling on that subject. Or, a variance from the expectations may just require a discussion between that member and you. In other cases, discipline may be required. It's only fair to address both expectations and consequences so every member has no surprises.

You have gotten through the meeting and you performed superbly. You stood your ground, explained things well and specifically, and made a few adjustments suggested by crew members that actually made those items better. There are some members who are not "all in" but you limited unproductive debate with respect. Now the next step begins.

From this point on it's your job to ensure that the expectations are met and that you meet them as well. You can't deviate from what you expect your members to do. Doing so will completely undermine the entire process and any credibility and respect will begin to erode.

> *"It's very hard to set expectations for your crew if you have none for yourself."*

As you start operating with your crew, you must follow through with the expectations you set for them. You can't let things slip by, even once. It is up to you to maintain the integrity of your expectations. Corrections need to be made and the more often you keep your members inside those boundaries, the easier it will become for them to stay there. In some cases they will test you, and you have to handle it right then.

You need to pay close attention and evaluate the progress and effectiveness or lack of both. It is really important to give it plenty of time. Members complaining is not a good measure of progress or effectiveness. However, not getting the results you anticipated over a specific allotted time could indicate that you should change something. It might be a little tweak that is really simple or it could require a trip back to the drawing board. Needing a change is not a sign of failure or an indication that your expectations were wrong or misguided; until you put them into practice, you weren't sure how some things would work out. It's okay. Don't let it get to you and don't let it change your resolve and confidence in what you're doing.

Failure to evaluate does your expectations and the efforts of your members an injustice. The whole idea is to meet the mission of the department, to be the best—so we must evaluate and make changes when needed.

On your timeline, set up additional meetings throughout the year to discuss progress or problems. A new meeting should also be held when a new member arrives. Sit down as a crew and go over the expectations again with everybody, not just the new member. Meeting again to see what is working and what isn't helps track progress too. Some members may have some very good insight and suggestions about how things have been going from their perspective. Value those comments, reestablish the expectations, and start again.

Expectations are important items—you can't just pay them lip service. Take your time, be specific, and set those boundaries. Pay attention and follow through and you and your crew will reap the benefits of clear, concise, and purposeful expectations.

Five Rules for Effective Expectations

Here are five rules you can apply to your members and crew in every meeting. Communicate these rules constantly because they encompass just about every expectation you will set.

Be productive

This is simple. Don't assign busy work. Don't assign work just because someone isn't doing anything or they have completed a task. Being productive means they take care of business, doing what needs to be done whether it's assigned or not, and contributing to the department's overall mission.

Be constructive

We want our officers and members to be constructive when mistakes are made or when others may be struggling with a task, assignment, or problem. We want a learning environment, and demeaning and making fun of others when they are vulnerable is not tolerated. We need our members to fix problems with respect and the dignity of the other member kept in mind. It's hard to learn when you are worried about how your peers are going to treat you.

Be generous

This is not necessarily monetary, although our captains have credit cards that they can use to help anybody who may need it. This is more about what members share with each member in terms of experience, knowledge, and mastery. Talk to each other about the calls you have run and the classes you have taken. Discuss the experiences with everyone, especially the mistakes you have made. We encourage the "war" stories from our senior members; we need to know about those calls. It certainly doesn't hurt for everyone to pick up a tab here and there, but that's not really what this rule is about.

Be professional

This encompasses who we are. How we answer the phone. How we answer the door when the public knocks. It's how we show up at public education events and public relations events. How do we look and act? Do we solve the problem with precision and expediency? Our demeanor after an incident should be one of tempered emotions while keeping in mind that somebody is suffering or upset at the same time. Professionalism is about who we are and who our community expects us to be.

Be aggressive

This encompasses everything on this list. We want members aggressive in trying to help in any way they can. We want our members aggressive in finding ways to have positive contacts in our community outside of incidents. We want our members aggressively training and drilling, learning to be the best. We also want our members to be aggressive on the fireground. We will aggressively extinguish fires, search for victims, and solve the caller's problems on service calls. Be aggressive to make a difference.

4

PLANNING AND PRIORITIZING

In this chapter we're going to start off with how to plan, prioritize, and schedule your daily activities, focusing on those with your crew, your shift, your engine house, or whatever unit you might be working in. It could be through a volunteer department, a paid department, or a combination department. All of these techniques and principles will apply no matter how you want to do it.

A common frustration of new company officers, whether a lieutenant or captain, is how to fit everything in with all that's already on the schedule. It can be very difficult depending on where you work, how often you work, and how busy you are at your particular department. When you start the plan, you have to get organized and that can be a real challenge for people. It's also hard to be consistent with getting organized and planning, but there are some basic steps that you can take for duty or in your personal life to get you more organized (fig. 4–1).

You need to develop and have some rules regarding your daily schedule. Whatever those rules might be, you have to be firm with your crew and with yourself about what's negotiable and what's not. You need to figure out early on, based on the expectations that you set, what is important and what are the things you have to get done.

After establishing those rules based on your organization's structure, you need to develop and try to set your schedule by the hour when you're on duty. That sounds hard, and it is, but it gives you a really solid, specific foundation to start with for how you're going to develop and create your priorities and when you're going to get things done.

Fig. 4–1. You have to have a general plan detailing when training and other tasks are to be done during the shift.

You must know and understand your organization's policies and practices. If your department says that at 8:30 a.m. you do housework, well then you have to plug that into your calendar. If it says at 2:00 p.m. you do physical fitness, you have to plug that into your calendar. It's best to break it down hour by hour. It doesn't mean that you're going to stick to that, but it makes you more aware and more cognizant of what you're doing and when you're doing it and how much time you have allotted throughout the day. Over time you will more closely know what is reasonable and what is not and you can make the appropriate changes. If you add five extra tasks to the schedule in one given day, when you start adding up how long it's going to take to do those five extra tasks, you'll find out quickly that you don't have enough hours in the day. Keeping a calendar, specifically by the hour or even 30 minutes depending on what you've allotted, forces you to provide yourself and your crew a more realistic list and plan to follow on a daily basis.

You may not get everything accomplished you wanted to. If that is the case, you're going to need to reevaluate your plan and your schedule with room for unexpected tasks. You've got to be adaptable. It can be very frustrating and it can be a challenge to feel like you're getting things accomplished. It's useful to have items that are expendable or ones that can be scratched off the list if something else gets added to your schedule. You're always going to have things that come up, from your organization, from the public, that are going to throw off what you're doing and what you're planning to do. You need to be adaptable and you need to plan on being flexible.

Additionally, don't sit down the day before to make out your plan. Look a week or two ahead at the rotation or schedule to figure out what's on the calendar. When you look a week or two ahead at what is on the schedule, just figure on there being two more things assigned that day by somebody other than you. This will help to prevent you from trying to schedule too many things on a given day.

You must be as consistent as possible. When you're setting your daily schedule, you're setting your plan. If you normally wash apparatus at 6 p.m. after dinner, do it that way every day. Don't make it 1:00 a.m. one day, 10:30 p.m. another day, 8:30 p.m. another day. Make it the same time every day. There are some things that can't be done that way. If you want to provide your members with an opportunity to work out and get some exercise, make it at the same time every day. It builds routine, it makes people comfortable and you'll see that things flow a little bit better. That doesn't mean it's set in stone. Because the public calls us for other needs, the chief is going to put things on your plate. But you should allow your crew to be as consistent as possible to help build a routine. Let's face it, everybody likes a routine. They like to be comfortable and they usually operate a little bit better in that fashion, so try to accommodate that.

When prioritizing tasks, you want to look at what's most important and what can wait. What are the objectives and are we meeting those objectives and expectations? When looking at what's important, you need to prioritize your day, what your mission is, what the organization's mission is, and then set your daily routine around that mission and vision. It should also be set around what your expectations were when you sat down with your crew. It might be that training is the priority.

Your chief has delegated to you their authority, but you're also serving the members of your crew, your shift, and your house. You're there to provide them resources. You're there to enable them to get the jobs done that you expect them to get done. You're there to make them better, to give them that knowledge and experience. You need to do your best to put everybody else above your own needs. That's what creates an organization and a culture of service and servant leadership.

Don't be the taskmaster. Don't be the one who has to do everything. Don't be the boss, the leader, the supervisor who feels like you have to do it all because no one else can do it as well as you. The reason you were put into a supervisory position, a management position, or a position of leadership is, hopefully, because you have the ability to motivate people to a) do things that they may not do on their own, and b) be more effective and efficient for the department, the company, and the citizens.

Planning and prioritizing is a critical step to being effective and efficient. Have patience and be deliberate in making your schedule. Don't get frustrated when the schedule gets blown up by a busy day; it should be expected and you should have a plan B to adapt to the situation. Don't underestimate the importance of having a routine and being organized.

5

SETTING GOALS

This chapter discusses ways to set goals and achieve them. There are many books, articles, blogs, and podcasts that claim to have the secret formula for setting and obtaining your goals. This book is not one of them. All I can do is share with you what I have learned. It is important to follow some kind of system or format to formally put a plan into motion to achieve your goals. Without a true plan, your goals and objectives will remain on paper or in your head without getting their day in the sun.

You first need to know what you want to do. Seeing your goals and realizing them are two very different things, and one doesn't always lead to the other. But with effort and persistence and a well-devised plan, you can reach your goals. Are there some limitations? Yes. The reality is that you cannot be anything you want to be. We have heard and read the exact opposite of that, but the truth is that not everybody can do everything; there are limits, many of them out of our control. But that should not discourage you. Most of the goals that we want for ourselves are very attainable!

Here's an example. You could wake up tomorrow and decide that your goal is to play in the NBA. Let's say that you are a 43-year-old, 6 ft 2 in. man who has never competed at a professional level in any type of athletics. At this age, you may be starting to feel and show a little of your age, but you're still in better shape than most of your peers. You can bench press over 220 pounds, can still run half marathons, and, as a firefighter, you have no problems throwing a pack on and working in limited visibility with the extra weight. You're still active and try hard to take care of yourself.

However (and some will argue this point), it doesn't matter how hard you work at playing in the NBA, the odds are very low that you will achieve this at this point in your life. The physical limitations alone will keep it from being a reality. You could practice every day and get a personal trainer, but it's still highly unlikely that you will ever be fast enough, strong enough, or skilled enough to make that happen.

The point of this is not to discourage you; it's to get you to focus. It is to get you to become intimate with what you want in your life so that you will set realistic goals the right way and achieve them, feeling the satisfaction of your sacrifice and hard work, rather than have the frustration of setting unachievable goals that will never happen (fig. 5–1).

Fig. 5–1. Establish attainable goals and work toward them.

Maybe you won't be an NBA player, but you could do some research and work toward playing in a charity game with some current or former NBA players. That is an achievable goal. It still takes work. It is still a challenge and will be a very rewarding achievement. Although the original goal is not attainable because of circumstances beyond your control, the second goal, which is related to the first, is a possibility.

SMART Goals

We need to consider the acronym of SMART for setting goals: Specific, Measurable, Attainable, Realistic, Timely. Specific applies not only to the goals but also to the path and objectives. Measurable is where some may struggle. How will you measure your progress? This is where you need to set some benchmarks that will be relevant to your plan and easy for you to understand and track your progress. Your goals need to be attainable. Think about the NBA story. They should be challenging and exciting, but they need to be realistic. Having a goal that you know will never happen is counterproductive and a waste of your time and the time of those helping you. You must have a timeline that is relative and reasonable for your plan. If it is too short you will be frustrated and more likely to give up. If you take too long, you will be more prone to procrastination and burning out.

Keeping the SMART system in mind, let's take a look at setting goals for personal and professional success. The steps can be used for you as an individual on a personal or professional level, or from an organizational perspective. Be sure to adapt and make it work for you and your specific situation. There are some steps you cannot alter or remove without losing the integrity and purpose of the process.

In order to get somewhere or to improve, you needed to have a firm understanding of who you are now. It is very hard to move ahead and direct yourself onto a path of improvement without having a good grasp of who you currently are and where you currently are in situation and status.

To gain a true understanding of what you want to achieve, you must have a mission. It should be what you want and how you want to be, not where you are or what you're doing now. Your mission should be a broad summation of where you want to be in life or professionally. But keep it short, precise, and to the point. It should not be a drawn-out document. For example, "I seek to positively influence those I come into contact with daily and to provide comfort and empathy to those who seek it out. I will be a person to whom others gravitate for advice, coaching, and honest discussion." The mission has to come first.

Second is the vision, how and why we are doing what we do. It may seem like the same thing as the mission statement. One way to think about these two starting points is that the mission—the who and what you are—for most organizations doesn't change. However, the vision—the why and how—might, especially with technology or changing demographics of customers.

On an individual basis, don't place a lot of emphasis on the vision statement. Some will argue you must have it, but I look at it like the mission is the spine of leadership. The goals are designed around getting there, so if somebody struggles with creating a vision statement, they tend to not do as well with the entire process as those who move on without it but have a clear alignment to the mission. You might have an aha moment as you are creating your core values and create a really nice vision statement.

After completing the mission statement and possibly a vision statement, it is important to determine your core values. This is especially critical for an organization. What are the building blocks that we put our credibility and integrity upon? What standards will we operate within to accomplish our goals and live our mission? These core values truly define your organization and you as a person, officer, manager, and leader. Honesty, empathy, sympathy, reliability, trustworthiness, caring, accountability, and many other characteristics and traits could be used as values. These things cannot be compromised during our daily lives and jobs or during our attempt at achieving our goals. These core values are at the heart of our process and what we will be defined by. From an organizational standpoint, direct, purposeful, and consistent disregard for these values will require strong action from leaders and officers.

This next step is where most people start: setting your goals! Yes, we finally got to this point, but you need to be methodical and judicious with your goals. A personal brainstorming session to write down all of the things you want to be, do, and achieve is a great place to start setting goals. There is nothing wrong with dreaming and it can make the process really fun.

Over time, start to get more realistic with your list and get rid of the unattainable and unrealistic goals. Those are roadblocks to you getting what you want out of life and your professional career. This is in no way telling you not to aim high; it is strongly encouraged to aim *really* high! Just keep those goals attainable and realistic.

As you get your list paired down to some real and exciting goals, you must start putting them in order of priority. Not every goal is going to happen in the same time period and not every goal is going to take the same amount of effort. What is needed for all of your goals is focus, attention, and consistency in achieving them.

You need to assign very specific objectives under each goal. A goal without a direct and specific path to achieve it is useless. This is an area that many struggle with. For example, a fire officer or potential fire officer may say that they want to take fire officer classes to be a better officer and leader. That is a very good thing to want to do. However, it is very broad and general and it does not give us a very good direction to move toward.

The importance of doing this for an organization and having key players involved in the process can't be stressed enough. This little tool can solve a lot of problems and keep both labor and management on the same page when indiscretions or differences arise. The answers can start with a simple question: "What is the mission?" Doing this with all of your members can play a huge role in keeping you focused and on track. It can also assist you in setting your priorities and staying aware of needs versus wants. Anytime there is a disagreement or conflict, you can refer back to your mission and goals to help make those decisions. It becomes less about who is right or wrong and more about what the mission is.

Having established your goals as an individual or an organization, list them in order of importance and chronology. If a goal can be reached much sooner than the other, it should be listed at the top of the list and attacked first. This doesn't keep you from starting on other goals; it just focuses you on the priority and helps you get a win as quickly as possible.

Now that you have your list fine-tuned and in front you and ready for action, you have to create specific objectives for each goal. It is not enough to just list goals and be general with the plan to achieve them. The more general and nonspecific the objectives, the harder it will be to stay on track to achieve the goals. If, for example, one of your goals is to take more fire officer development classes, you need to break it down into exactly which classes or topics and set a timeline for when you will attend and complete each class. Simply stating that you will take more classes is not specific enough and does not create an actionable plan. It doesn't commit you to moving toward achieving that goal.

There are times that circumstances out of your control will derail your best plans, causing frustration and disappointment for not achieving your goals on time. That's not the real issue though; the problem comes when you don't reset and change your objectives and timelines. You have to just start again. If coming up short is truly from a situation you cannot control, that is not a failure. It is simply a setback, one that you can overcome easily.

6

DELEGATION

New company officers—in fact, any new fire officers—can quickly become overwhelmed by the daily responsibilities of their new position. It's easy for these new officers to underestimate the demands placed on them, and they are eager to succeed. With that eagerness they forget that they have some help in their crew and try to do everything on their own. This causes several problems for the new officer and their relationships with both their crew and their superior officer.

One important thing to understand is that delegation is a show of trust (fig. 6–1). Being effective as an officer requires building relationships with your members and creating trust. Fire officers who struggle with delegation run into two major problems: 1) They become overburdened from wanting to do everything on their own, and 2) by not delegating any tasks or assignments to the crew and members, the new officer appears to lack trust in them. You have to let go and share some of your responsibilities.

Fig. 6–1. Delegation is a show of trust.

Delegating allows you, your members, and the crew as a whole to be more effective. Tasks and assignments are more likely to be completed in a timely manner. When the officer has a lot on their plate and needs to meet some deadlines, one of the best ways of completing a task is to delegate it. Delegation allows your members to complete smaller tasks and ones that don't need your full attention, giving you more time and focus for more important or complicated tasks that only you can do.

Delegating some of the smaller tasks and assignments to crew members has the added bonus of those tasks receiving more attention. When your members are given a job, they can give more attention to it than you might have been able to provide because of your other responsibilities. This is not to say that you wouldn't give it the proper attention, just that a member who is getting a special opportunity to engage and make executive decisions may give it a priority level you wouldn't. Offering and assigning tasks to members and the crew is an effective way to create some ownership. When they are responsible for the outcome of a task or assignment, they are more likely to take ownership of the project, and that could help build their ownership and accountability in other mission areas of the department or organization. It allows and encourages members to feel like they are a key part of the crew, the department, and the mission.

When done correctly, delegation by an officer to a member should include an open line of communication. The act of delegation should be an opportunity to improve communication skills with your members. In most cases, it becomes a simple act of providing precise instructions and then listening from that point on. It is a great chance to listen to your members' needs and provide the resources requested. It is also the perfect time, if appropriate, for budget requests and summaries to be written and submitted by the member to the officer, which can be used as opportunity to train potential officers.

You might find a diamond in the rough when delegating. We aren't always aware of some of the talents our members have. Putting them in a situation that helps them showcase what they're good at empowers them to do more and to share it with others. Uncovering the skills and talents of our members helps to instill confidence and pride in that member and to understand and better know what your members are capable of. You never know when the need will arise for those specific talents and skills.

One of the most beneficial outcomes from delegating to your members is that the successes are theirs. They get the accolades and satisfaction of successfully completing the assignment, no matter how large or small. This provides reinforcement of the fact that they are owners of the mission and a key component of the department's successes.

When members are delegated tasks and assignments, they have an opportunity establish themselves as reliable and to better understand the core mission of the department. In the process of completing delegated tasks, they can learn how to better gather and utilize resources, both inside the department and outside of the organization. It can highlight their ability to be resourceful when completing difficult tasks.

Some members who have not been so engaged in recent weeks, months, or even years may have an awakening from a delegated task. Most new officers won't ask these members to take on a job or assignment. You have to be brave enough to ask and show confidence in this member to make it fly. It may take several attempts to delegate to these members to get them to accept, but their success just might be enough to bring them back around to being productive.

Delegation can be a great tool for mentoring and coaching personnel for development. If you have members who want to be or are potential officer candidates, you can use delegating as a tool for sounding out shortcomings and unknown strengths. Assigning very focused and specific tasks that have a purpose can assist in pushing or challenging a member to rise above their normal limits. Just as it can be a great tool for evaluating up-and-coming talent in your department, it can also be a great tool in encouraging dormant talent. As a new fire officer you may be part of a management team that is making some unpopular changes. You might not be seeing the buy-in that is expected. Carefully delegating tasks that are part of the mission can help push change from the bottom up. If done with care and diligence, assigning mission-based tasks to members who may not fully be convinced can be very helpful in changing minds and attitudes. The change doesn't come from the delegation of the task, but from the research done by the member.

How to Delegate: The Process

There is a formal process you should use to effectively delegate and to receive positive results from your members. You can adapt these to better fit your department and situation, but this process is simple and not specific to any one kind of organization or position.

Plan

Giving orders at random is not delegating. Delegation should be planned and thought about before offering the task or assignment to the member. There needs to be forethought and deliberation on your part before you just throw out a job to a person on your crew. You need to consider who is best suited to complete the task, how you will present the job or task so that it is acceptable to the member, and how it fits the mission of the department. You can't shoot from the hip—the member will struggle and you will have a difficult time ensuring you are doing all you can to help them succeed.

Train your members

It will be difficult for the even best of members to complete tasks assigned to them if they haven't been trained and educated appropriately. There has to be a correlation between a firefighter's training and the assignments they are delegated. Whether it's research to buy new fire hose or teaching a new skill, there must be some basic, fundamental training on the subject. For example, if you assign members to research and purchase fire hose, they will need to understand the flows your department wants to achieve, how it is loaded, what type of nozzles are utilized, and many other specifications. They must understand the importance of fire hose and how they operate the fire hose.

Establish boundaries

This step is one of the most important in the process. For fairness to your members, when you delegate a task you must establish the ground rules. This is the conversation that provides the timeline of the project, the frequency of progress reports, and the expectations of necessary resources. These have to be clear and concise and the member should ask any questions they might have. Boundaries are critical to the success of delegation.

These boundaries are not just for the member you delegated. You need to adhere to them as well. If you are going to set boundaries and set deadlines, you must follow them. If you say you will provide resources, you had better be quick to get them. If you set time intervals (one week, two weeks, etc.) don't put your nose in the project until that time interval comes around. When you get too heavily involved, you violate the terms of your agreement and become a micromanager.

Clearly define the desired outcome(s)

The same way we need to set boundaries, we must clearly state and define what needs to be accomplished. It is our responsibility to provide a specific explanation of what the endgame is. Our members who accept delegated tasks must know what they need to do and where the finish line is. The finish line shouldn't be moved or changed without a discussion with the member. It should be a very rare instance that the final outcome is changed. Additionally, the benchmarks and boundaries must stay consistent throughout the process. If delegation in large part is about establishing trust, we must remain consistent in what we expect and in the boundaries we expect our members to operate within. Not doing so will create integrity issues for you as the officer.

Review and evaluate progress and results

After setting the boundaries and establishing what needs to be accomplished, we need to ensure that we put into place the ability to monitor the progress of the assignment. We must define how this will be done during our conversation on setting boundaries. There has to be some oversight on your part to ensure that things are on point and to offer any assistance if trouble is brewing.

In some cases, you may have to gently and subtly intervene to ensure things stay on track. But this should only happen at the designated times established in the boundaries and expectations. Ask if you can help, don't just take over the project. A better approach is to offer additional resources or offer the assistance of somebody you know can help them who isn't you. The key is to allow your members to play a large role in the decision-making process.

During this review and evaluation period, don't change course unless there are no other options. That decision is best made by the member, not you, after exhausting all options for completing it as assigned. If the members make the change, they still own it, which is what we want. If we make the change, especially with no input or consideration from the member, they no longer have ownership and the delegation is probably over. Try to avoid that situation.

Plan B

Sometimes the best option is to let your member fail. The best lessons are learned from frustrations, challenges, and failures. When a member has put their absolute best foot forward and they worked hard and things just aren't working out, failure may be the only outcome that brings to light where they need to improve. If you allow them to fail, be sure it is done in a productive and constructive manner. It should be a learning experience, not a humiliation, and it should not determine whether or not you assign them future tasks.

Since we know shortcomings happen, it is wise to have a plan B in place before you delegate that specific task. This is part of the first step when you plan your delegation. If it is a project that has to be completed and we choose to delegate, it is important to have a Plan B in the event that the delegation doesn't go well. It may be because the member gets sick or injured, leaves the department, or just doesn't complete it. Either way, a plan B will allow you to pick up the pieces and finish it. In many cases you can implement plan B with the assistance of the member who didn't complete the task. Doing this, you keep the member engaged and make it a learning experience.

Delegation is a critical component of leadership for new and veteran fire officers. You don't have to, nor should you, do it all on your own. Learn your members and assign tasks and jobs that will help them be successful and take some of the load off of you.

7

HANDLING DIFFICULT MEMBERS

Being appointed our department's first training officer in several years and being tasked with creating a formal training program at our well-established fire district, I knew that I was going to face challenges from many directions. Not everybody on our department was happy about the idea of making training a consistent part of our day from now on. That, I expected. But I was not fully prepared for the attempts from some to completely blow the whole idea out of the water.

Half wanted the training and the opportunity to hone our craft and the other half did not, for multiple reasons. The ones who did not want training were not unified by a cause; they all had their own selfish reasons for not wanting to be "made" to train regularly (fig 7–1).

We heard all the excuses: we're too busy; we do a good job now; what can I learn, I've been on the job for almost 30 years; and so on. Some just did not want to have to show that they could do all they claimed to do and others did not want to be inconvenienced. Regardless, after full consideration of the change we were implementing, we went full steam ahead with tempered optimism.

I started to implement more hands-on training and evaluate performance. I also started to bring in outside speakers who were nationally known. I decided to bring in three a year, to make them very cost effective for outside fire agencies to send their people to, and to make each of our shifts attend one per year on duty. It was an exciting time for our department—or at least for half of us.

As we proceeded with our program and it became more clear where people stood with their opinions of training, it became more important not to participate in the chatter. We focused on finding ways to bring in the members who were not convinced or happy about a more intense and consistent training program.

Fig. 7–1. Those who want to train will show up to the drills; it's the ones who grumble and complain that you have to worry about.

The one thing that we didn't do was let the opinions and thoughts of others affect our opinion of anybody. We gave each member a clean slate and went to work at earning their respect, not demanding it. We started to build relationships, which turned into trust. We treated each resistant person as a member who had knowledge and experience to share and learn from, not the next person who should retire.

Set Expectations

As mentioned in chapter 5, setting expectations lays out the plan and boundaries for your members and yourself so that there is no question about what is acceptable and what is not in regards to performance and accountability. Where many go wrong in this aspect is the follow-through, especially with difficult or aggressive members.

Loud and aggressive people often get their way. Nobody likes confrontation, especially at work with people you live with for a third of your life. Additionally, most people want to be liked. In order to keep the aggressive and loud member in check, you will have to hit their attack head on, stand your ground, and make at least one member mad—the aggressive and loud person!

What gets lost in considering this confrontation is that, sure, you make one person mad, but if you don't deal with them up front and in a timely manner, you will make more than one person mad by not stopping the troublemaker. It simply is not worth it to be passive. We have to quickly set boundaries and stick with them. They must be clearly communicated and understood early on in the working relationship.

These expectations and boundaries must be specific and, for these members, may need to be reiterated more regularly than what you would like, at least in the beginning. They may need subtle, or not-so-subtle, reminders of what you expect. When they aren't meeting your expectations, you have to let them know and correct the issue.

You may need to be a little more patient than you'd like because they are trying to break a habit of being miserable, so to speak. It will not be an easy transition. This does not mean you make an excuse for the member, only that you have to have a realistic timetable for the changes to take hold.

In most cases members who are perceived as difficult or obstinate no longer feel valued in the organization. For whatever reasons (and they may be legitimate), they have lost touch with the mission and no longer feel like an insider, but rather a person who comes in and is told what to do. They will find everything wrong with anything and are just looking for a fight because that is what they feel is expected of them.

As the officer, you need to find their value. Everybody has it and it usually isn't that hard to identify. It's up to you to figure out how to bring it out and to expose it in a positive manner.

I had a senior member who had by all accounts been labeled. Most were ready for him to retire. He could be obstinate and difficult to motivate, but he was borderline genius and nobody else in our department could explain, in such great detail, how an engine pumped water. This guy could draw every intake, valve, and elbow of the pump and the piping in every truck we had. But he had no interest in teaching because of how he had been treated in the past. Now, he carried some of the responsibility as well, but many just didn't value what he provided to our department.

We started by putting new drivers and operators with him to go over the truck checks. Then we asked him to train them on the pumps and ladder truck. Let me tell you, nobody was as thorough as he was in explaining and showing how everything worked and he talked in a way that was easy to understand. What he lacked in patience with some, he made up for in sheer knowledge. And in time he improved in patience too. Eventually, when I gave him new drivers and operators, I just walked away.

That was the key. I never looked over his shoulder and I always trusted what he told me about the progress of each candidate. Unless proven otherwise, I accepted what he told me as the gospel. I had to show him that he was valuable and that I trusted him. There was no way for me to do that if I was constantly second-guessing him during the process. The result was a more engaged member who many had thought a lost cause and excellent junior drivers and operators.

Dealing with difficult members takes patience and a direct approach. We can't treat them any differently than we do anybody else. We must have an awareness that each person has a value to our department that may have been lost or tucked away because it was never appreciated to begin with.

For others, those more angry than injured, you must hold a stern conversation with real consequences. Not every member is going to be responsive to the examples that I shared. Not every member is going to be easily led to becoming engaged again. For those members, you will have to establish those boundaries and expectations, ensure that they are communicated so that the members understand them (not necessarily agree with them), and provide consequences when those expectations are not met and the boundaries are violated.

In those instances, you can't be afraid to use discipline to correct behavior. A huge problem with newer officers is the avoidance of discipline (see the final section of chapter 2). In an attempt to stay friendly or to avoid an awkward situation, they go easy on members and give multiple chances to change. But doing this enables those behaviors, diminishes your credibility, and ultimately leads to the loss of your crew or shift.

Discipline is not a bad thing when it is used correctly and in a fair and consistent manner. What we have to remember is that it is there for a reason: to change behavior. We are doing our members an injustice if we don't provide them honest feedback. Sometimes, that feedback is a disciplinary action.

Our job is to keep the order and to develop our members. If we can't hold them accountable, we lose our credibility and ability to lead. Most situations can be corrected with a simple conversation, but there are other times that discipline must be used to not only correct the behavior of the offender, but to let everyone else know that when they don't follow the rules, there are consequences.

This will be especially difficult with friends and members with whom you have spent a great deal of time. But don't be persuaded to avoid these confrontations or the use of discipline. That decision could make or break your reputation as an officer and result in a loss of trust from other members when you let certain people, especially friends, get by with things that others wouldn't be allowed to.

Handling difficult members and handing out discipline is never easy. But with an even temperament and patience, you can be effective in dealing with those situations. Remember that avoidance is as bad as overreacting and neither should be an option in dealing with these members. Have an open mind, don't rely on perceptions, build trust, show you value everyone, and create relationships to help overcome some of these challenges.

8

HUMILITY

We are all going to make mistakes as a new or current officer, it's just part of being human. Being able to own up to them is an area that takes real humility and courage. We've all had weak moments during our careers and during our lives and in most cases, if we're paying attention, they taught us valuable lessons about life. When you tell your children what not to do as they're growing up, in the back of your mind all you can think about is when you did those same things.

We will be tested and watched to see if we are walking the walk and taking our own advice. It is not hypocritical of you to make corrections to others who need it. That's our job as officers and leaders. The real key is how we do it and how we address our own mistakes.

As you advance in your career, as soon as you try to make a correction, try to change someone's behavior, don't condone a conversation or tone, or sit down and set out some expectations, somebody or a group of people is going to immediately question you: "Who are you? I remember when you did these things." "Oh, you're the guy who said these things." You know what? Don't hide. For us leaders, it starts with having some humility. For us to deny those comments completely is a huge mistake because it's going to set the tone for how you respond in the future. We have to openly admit when we have made mistakes in the past. I like to say this to people who call me out on some of these things: "Yes, you're right, but I've changed. I did some pretty stupid shit in high school that I never want my kids to do too." The point is we grow and learn.

It also shows your ability as a leader to say, "Yes, I did do some of those things and I was wrong. I have had those issues and no, I wasn't always a perfect employee and I'd be ridiculous and naive to believe that or try to make you believe that, but I evolved. I changed. I grew. I matured." All of us, when we were in our early to midtwenties (and some people in their early thirties), made some really stupid mistakes and said some really silly things because we hadn't experienced the things that those who had been around a bit longer had. It's just a fact of life. That's why I should listen to my 75-year-old uncle when he talks, because he can impart a lot of wisdom to me about things he did in his forties and fifties (fig. 8–1). It's the same with my dad and the people a generation older than me.

Fig. 8–1. There is always room to learn something new.

We have to have some humility to listen to somebody offer criticism and tell us all of the things we do wrong, then take stock of that and not take it personally. There's nothing more powerful for the people who look up to you, the people you supervise, manage, and lead, than for you to say, "Yeah you're right, I did do those things, but you know what? I got past it. I corrected the behavior. I don't do that anymore and I haven't in a long time. It can't be condoned and we don't want those mistakes made by other members of our department."

Having humility also means that you accept that what you did or are doing is a problem. You have to accept that you screwed up. You have to accept that it was your fault and that's not an easy thing for anybody to do, whether it's professional or personal. It's never fun to do that.

Just consider how many people can't say they're sorry; sometimes I'm one of those guys. I'm human. We all are and we have those challenges, but as a leader and as a manager and as a supervisor, we have to own our mistakes. It's called accountability. If we aren't accountable and don't take responsibility for our actions, how can we require our members to be? We'll never get any better if we don't own our mistakes and realize that adjustments need to be made. It is very difficult for us to help others improve when we can't do the same for ourselves in an honest way. As uncomfortable as it can be to have your crew or members point out what you're doing wrong, it must be conveyed that they have the latitude to do so, respectfully, without retribution.

Humility has to come into play when you make mistakes that your people might see but don't speak up about because it hasn't been clearly communicated that they should or there is a lack of trust. We had a call like that one day when we were operating in a spare apparatus. At that time the computer system didn't connect to the spare truck so we weren't dispatched to a fire in our immediate response area. The house tones did go off for the chief, but I recognized the address.

I was slicing cucumbers for dinner and I started listening to the address and the cross street and I was like, "That's our fire," so right away I was feeling behind the eight ball. I was already rushing. I was already anxious, already thinking "We're going to be late to our own fire that they didn't even dispatch us to." I rounded up the crew and let them know that we needed to respond to this call.

Anybody who has worked on a fire apparatus knows that there's not a lot of room to put your gear when you're not going to a fire. It's always on the doghouse or somewhere and so I would put mine on the side compartment behind the backstep firefighter. I went to the compartment and grabbed my gear and was working to get out the door pretty quickly. While we were in the middle of the intersection, I realized that I'd left the compartment door open when I got my gear—totally my fault. We had to stop the truck, get out, close the compartment door, and get back to trying to figure out exactly where we were going. I knew the apartment complex that we were responding to, but I didn't know the exact address or street since we hadn't been dispatched to it initially.

When we pulled into the apartment complex, I had a maintenance worker waving me in. I knew in my heart that wasn't where we should be going, but I thought, "Hey, this guy works for the complex. He knows this place better than anybody. I'm going to do what he says," which is something I never do. That was completely out of character for me.

Now we were on the back side of the building in the wrong place. I gave an early size-up and it was so fast and rushed that it provided very little valuable information for incoming crews. Additionally, I didn't take the normal tools that I would typically take, again, because I was rushed. One thing I've always prided myself on is preparation and being pretty levelheaded. When you read line-of-duty-death reports on firefighters, you hear a lot of people talk about how the dominoes fall. It's not one large, catastrophic event most times. Many times it's a lot of smaller, seemingly insignificant events that lead up to one problem, which is a person perishing or being badly injured. This could easily have been one of *those* fires.

Lucky for us, it was a blower motor in the basement and not an actual fire. I remember getting back on the truck feeling very embarrassed and very frustrated as a man who has spent a lot of time in training and developing course work. There were plenty of people who would love to jump on that because not everybody wants to see you succeed or do well.

I could have gotten back on the truck and just put us in service and driven back to the firehouse and never said a word about it. It never would have been talked about. Maybe among themselves, but never with me. They never would have gotten the validation of my mistake and may have acted upon it in the future. But instead I got back on the truck and I turned around and told my crew, "Listen, if that ever happens again, if you ever see us operating out of the routine because of me or anybody else, everybody has to speak up. Everybody has the ability to tell me or whoever else is in that front seat to stop and take a look at what's going down because if they're wrong, they're wrong. If I'm wrong, I'm wrong."

We had a really nice conversation about how they could definitely see that we weren't operating the way we normally operate. Also, handling this in an educational way made for constructive conversation and discussion. It also gave them formal permission to speak up when something isn't going right.

That's the humility you must have as a leader and supervisor. Not only do you have to give permission for that criticism and correction to be spoken, you must also accept being corrected by others and take constructive criticism offered by people around you, those you like and those you don't. You must understand that they are seeing some legitimate issues and are trying to offer you what you need to act on them.

More importantly, you must be able to admit your mistakes, take action on them, and use the experience to become better. The worst thing you can do is get an ego that is too big or a pride that is too full. It will be a frustrating career and life if you don't allow yourself to open up and be a little bit vulnerable and humble and to accept the words and advice of others. As leaders, if you're going to try to improve, mentor, and coach others, it's critical that you have the ability to be coached, influenced, and mentored not just by those above you but also those below you in the rank.

Humility is not just about when somebody tells you, "Good job," and you say, "Aw shucks, thanks, it was really a team effort." Those things are important as well; you want to give credit where credit is due and not be too boastful. But humility most importantly should be used to admit mistakes, take ownership, make the corrections, and make the right changes to better everybody, including yourself.

So be humble, get rid of the pride that holds you back, and listen to the people who work with you and for you and empower them to make those corrections. Empower them to speak up. Empower them to give you constructive feedback even when you don't ask for it. It's a lesson I've learned the hard way in my personal and professional life as I'm sure many of you out there have in all walks of life. You're going to be faced with challenges and it's going to be very uncomfortable when people call you out, but try not to take it personally.

9

LEADERSHIP: DO YOU HAVE IT?

Are You Ready to Lead?

Many like the idea of being promoted and being in charge of a crew or company. They like the idea of having a white shirt and some gold brass on it. But when it comes time to deal with the responsibilities that aren't quite so popular or easy, they just stick their head in the sand. We have to be confident and positive that we are ready to lead in all aspects of what the company officer is responsible for.

Just putting yourself out there for the promotion process requires a degree of risk that some won't take even though they have the tools and skills to be good officers. They are not willing to put themselves in a position to fail in front of others. That in and of itself is an indicator that they are not ready to take the lead. You must be ready to make some hard decisions that sometimes involve longtime friends and crewmembers. When you get promoted, your job changes and you must understand that your responsibilities are new and different (fig. 9–1). You will have to say no and the sooner you're able to do it, the better off you'll be. You will have to be decisive and firm. In many instances, this kind of risk is unexpected and that is something that new company officers struggle with the most.

Just like on the fireground, risk aversion can have very dangerous outcomes for firefighters and for victims. Risk avoidance in the firehouse can ruin the career of a new fire officer.

Fig. 9–1. Your responsibilities are different now.

Why Are You Here?

It's a simple enough question, but one that is critical to what we do and how we do it. The truth is that most of our job is not doing what we dreamed or expected it would be. At a young age we all thought we would be making a difference in a big way every day. We envisioned ourselves running into burning buildings on a regular basis, cutting people out of cars, and bringing patients back from the dead on most ambulance runs. That just isn't how it works.

Most of our job is still about making a difference, it's just in different, but no less important ways. You should be here to do the little things every day that make a huge difference to others, not necessarily fulfill your youthful dream of how the fire service is or should be. The fire service is just that: a service. As leaders and as new officers you must understand why you are here—to serve.

That brings us to the next question: Why do you want to get promoted? It's one that you will likely be asked in any interview for promotion, and people typically provide a generic response. You must understand exactly what your role should be as the formal leader and officer.

It's Not about You!

I was at a shop meeting many years ago when the discussion turned to the actions of one member who hadn't been doing things the right way and whose career had started a downward spiral. He had been circumventing the system to get where he wanted to go without the work or qualifications. He just couldn't wait and do it the right way, it took too much effort. This member went from being popular and well-liked to losing his influence very quickly. Over time, he was promoted to captain under very controversial circumstances and more influence and respect were lost. A short time later he had an assistant chief position created for him and he promoted up with no process. That was the end of his career.

When the other members rallied to bring integrity and honor back to our process, they couldn't be stopped. The political board members changed, the expectations changed, and he left a career way too early on his own after taking several months of sick leave.

Right before he left, he spoke to me privately and expressed his confusion. His estimation of the situation was that yes, he didn't get his position in the traditional manner, but he bought whatever the members wanted, expecting the ill will to blow over. It never did.

His focus was always on himself, never on the organization and the people that he served. He thought and believed that his promotion was what gave him the influence and authority to *make* members respect him. He didn't understand that his most important job was not to look out for himself; it was to *serve* everyone below him in rank and status. By putting the emphasis on what the increase in rank could do for his image, he blinded himself to what the real job was and didn't see that any influence and respect must be earned through selflessness.

Concern and Empathy

An area that is not always discussed enough when talking about leadership is having genuine concern and empathy for your members and for those you serve in your community. It's easy to get wrapped up in whatever is going on with you or in your tight circle of friends (see the section above). Being an effective and respected leader requires you to show some compassion and concern for others.

It can be difficult for your members to show concern and empathy for others if you aren't explicitly expecting and requiring it as well as demonstrating it yourself. Your actions speak as loud as or louder than your words. If you are unwilling to have a soft side in regard to trying to understand and react in kind ways toward those who are struggling, you will miss out on the chances to build the relationships necessary to be an effective leader.

Your kind words and show of concern may be more important than you know for that person. With the high rate of suicide in the fire service, we must try to relate and be compassionate toward others. We don't always know what is going on in someone else's life or what they may be struggling with. If we just write them off as being soft or some other stereotypical reference, we might miss a critical warning sign.

Put Yourself in Harm's Way

I know what you're thinking and that's not what I'm talking about. I'm not talking about on the fireground, although that may need attention as well. No, I'm talking about in the firehouse. It's important for your crew to trust you and to know you have their back at all times.

There is nothing worse than an officer who will pass blame down to their members when difficult times come around. This is especially true when tasks aren't completed or a problem has been relayed to you about your members and or company. The last thing you want to say is "Well, it was Joe, he's just horrible," or something to that effect.

You need to take responsibility; it's your member or company. It goes a long way in creating and maintaining trust. It also says that you are not going to pass the buck and join in on blaming or singling out a specific person in front of everybody else.

This does not mean you turn a blind eye; quite the contrary. What it means is that you handle problems in your own house. It is something you correct with that company or that person after *you* have taken responsibility for it. You're there to shoulder the responsibility and that means taking the heat and then passing it on as needed to make the proper corrections.

Don't Be a Screamer

Getting louder and more animated does not mean people will better understand what you are trying to communicate. In fact, the more you scream and shake the less and less they will listen. They become numb to your tantrums and tirades. Soon, even if you're right, they no longer hear you.

Additionally, you just look and sound ridiculous. There have been very few instances in my experience where increased volume in a message was necessary, well received, or constructive. You cannot scream your way to success and you cannot build relationships when the only way you know how to handle adversity or frustration is by screaming.

Instead of screaming, think long and hard about what you're about to fly off the handle about. Be patient and figure out a way to do it in a normal tone of voice and in a way that is constructive. If there is a true problem, you want to solve it, not scream it into submission.

Demeaning and Degrading

This goes right along with the screamer, because in most instances the person who provokes you to one is the same person who provokes you to the other. If you demean, make fun of, or degrade fellow firefighters for any reason, you are not officer material. Actually, you're not firefighter material. It really is cowardice to purposefully put somebody down in a negative, hurtful manner. This normally comes from a person who is not very secure and is pushing attention to somebody else to avoid having others see their own faults and missteps. If you are reading this and asking yourself if you do this, you probably do and you need to stop!

Our job, especially as officers and leaders, is to lift members up. It is our mission to do everything in our power to be a positive influence. As the officer, when you do see this taking place, you have an obligation to stop it *now*! It cannot be tolerated or accepted. Walking past it condones it and is as good as saying, "Go ahead, keep doing what you're doing." That simply is not acceptable.

True leadership exists when people follow you no matter what your rank or title is.

Do and Say the Small Things

You have to figure out what is important to your members and what is going to make them feel good about coming to work. You need to pay attention to the things you can do and say that will make their job easier. No, you aren't doing their job for them, but some actions and words can make the assignments you've given them easier to finish. Just having a hot meal when they get back from a run can mean a great deal.

When I was first promoted to captain I had a good friend and fellow firefighter on my shift offer to give me some advice if I was interested. I respected this guy and I was very curious about what he had to say. He told me that there is nothing worse for a paramedic on the ambulance to get toned out during dinner only to come back to a cold plate and dishes that needed to be done. Our crew would never leave dishes for the ambulance crew, but apparently some did. (However, I *was* guilty of not taking the crew's plates and keeping them warm in the oven while they were on the call.)

That was one of my first expectations of my new crew when I moved shifts and it was noticed. The ambulance crews really appreciated that gesture and it made running that call in the middle of dinner just a little bit easier to deal with.

Tell your crew "thank you" when they do things right. We spend so much time and effort looking for deficiencies or what people are doing wrong that we forget that most of the time our members are doing things right and going above what is expected. It's good to acknowledge that. It must be genuine and not overbearing, but if you pay close attention you'll see the right times.

Occasionally tell your crew after an alarm sounding call when they were all buttoned up and with tools in hand, "Thanks for being prepared." I know it's their job, but you want it to be reinforced. Pick those small, easy times to just say thanks. Tell people good morning and say hello and all of the simple little things. It makes a difference.

Being Honest—Easier Said than Done

On the surface, being honest looks like an easy issue to deal with. In most cases I would say that is true. But being honest can be a complex thing. Sometimes a person who is struggling has to be told that the problems are not outside factors but them. Filling out performance reviews or evaluations are another area where being honest has its downside. Nobody likes to hear they are deficient and why.

Being honest with day-to-day things is usually a piece of cake. These are situations when people expect to be told the truth and there is little pain in hearing it. An example might be that a member didn't pass an exam or that someone else was promoted. It's when the small specifics start getting brought up that things can get hard. "Why didn't I get promoted?" "Where do I need to improve?" Questions like that can be incredibly hard conversations if we're honest. Having to tell a member that they can't teach a class because they're just not that good is not going to be a comfortable conversation. Being asked, "Do you think I will ever get promoted?" from someone who just doesn't have the tools is a rough conversation.

No matter what the subject and topic, you must be honest. Not brutally honest necessarily; you can choose the right time and words to make it as easy as possible, yet still on point and true. Just keep in mind how you would want somebody telling you or a family member they didn't make the cut or weren't good enough at something to advance. You have to find that balance between the feelings of others and still being clear.

What is not an option is avoidance. You cannot put off these conversations. It's not fair to the person you're talking to and it's not fair to you. Tough issues must be dealt with by being methodical, just, and fair, in a timely manner that doesn't force the firefighter to wait too long for an honest answer. Honesty really is the best policy, just be appropriate and timely with it.

Your Moral Compass

True leaders operate under a code. It consists of integrity, pride, and values that are not up for negotiation. The decisions that are made, both easy and especially difficult, are based on the values and code a person carries with them. It is your moral compass, your interpretation of what is right and wrong.

A person with a poor values system is going to struggle as a fire officer. There is just too much that the fire service stands for that is based on basic values and ethics, and when these are deviated from, it is obvious that the person should not be trusted. Once there is a lack of trust, leading is almost impossible. Sure, they can order others around, but they won't be leading them.

If you have low morals at home, you will have them at work. If you don't value others and the principles that the fire service was built upon, you're going to struggle to identify when others are not following them, putting our profession and departments at risk.

Tactics can change, leadership styles can differ, training drills and evolutions can vary. But your core values have to be rock solid and unshakable. They cannot be broken or compromised. This may put you into some really tough positions with other firefighters, other fire officers, and friends, but you have to maintain your core meaning for being in this profession and position. Once you compromise your principles, you start down the slippery slope of not knowing what you will and will not stand for.

This boils down to honor and integrity. Doing the right things, for the right reasons, the right way, all of the time is integrity. What you do when nobody is looking is who you really are and defines your honor and ethics. Don't let either one be blemished.

Your leadership is your leadership. You really have to adapt it to your members so that you get the most out of them. If you try to lead based solely on what your own idea of leadership means, you will not be as successful as you could be. You must serve your members, lead so that they follow, and adapt to make their experiences in this profession the best they can be.

10

COMPANY DRILLS FOR THE COMPANY OFFICER

An area that is a constant in conversations with newly promoted company officers and veteran officers who are trying to change is company drills. "How do I do company drills?" "What material should I be using?" "How do I know what topics I should be covering?" "With a busy schedule, how do we fit them in?" "Members on my crew don't like to drill, how do I motivate them?" These are just a few of the questions many company officers have regarding drilling (fig. 10–1).

Fig. 10–1. Motivating the crew to drill can be a challenge at times. (Metro West Fire Protection District staff photo.)

This chapter breaks down some simple steps to assist you with being more confident in delivering company drills on shift or at your volunteer house. However, you must understand, there is no silver bullet when it comes to drilling and motivating your crew. You have to get out and get dirty. There must be constant effort on your part to keep your members motivated and informed about the why of each skill, task, and drill.

The options listed are suggestions based on real experience. They may need to be adjusted for your members, your crew, and your department. But the basics are the same no matter if you're paid, combination, or volunteer: get out and drill on the fundamentals with purpose and engagement.

As a new company officer the first few weeks are filled with excitement and hope—hope of making meaningful changes and implementing an attitude of engagement and proficiency. It's exciting because you have the opportunity to really make a difference in your department and to bring all of the things you have learned from classes, conferences, books, blogs, and podcasts to your members. If only it were that easy.

With the excitement and vigor comes the tendency to rush and be in a hurry to do things. One of those things is to drill at the company level. Starting right away is not a bad thing, you just need to have a plan to use so you are drilling on things that you are not so good at, addressing expectations, ensuring that you are operating and drilling on your department's standard operating guidelines, and making sure you have the ability to adapt to changing fireground conditions.

Without some organization and forethought, drilling becomes boring, inefficient, and inconsistent. Or it becomes the same group of drills over and over because they are familiar, fast, and easy. That is a recipe for failure and one that you want to work hard to avoid. The good news is that creating, implementing, and delivering company drills is not that hard and you can get really good at it quickly if you use your resources and are patient.

Identify Needs

One of your jobs is to determine what you and your crew need to get better at. It is critical to identify the areas you are good at and the areas that need attention. In most cases it is not that difficult, but you can easily overlook areas that are more cumbersome to change, especially if you don't have expertise in that particular area.

Identifying needs must be done through activity. There is nothing worse than to sit your crewmembers down and make them take a written test on the basics of firefighting. Many officers and training officers like to call this a needs assessment. I don't agree. A written test will measure retention and the ability to recall specific pieces of information, but it does not measure the member's ability to perform tactically on the fireground.

Written needs assessments can be valuable for a training officer or chief to establish a baseline for an entire department in creating curriculum or to meet accreditation standards, but to use this for company drills is counterproductive and does not provide the desired information for improvement on the fireground.

If you want to figure out if your members are cognitively sharp, have frequent sit-downs at the kitchen table or front bumper and discuss topics, articles, chapters from text books, line-of-duty-death reports, and other materials that are related to the activity you want to address. The benefit of this is that you can do it every day of your shift or every training night, not just when taking a test. Additionally, you can send out the materials ahead of time via email and have your crew view them before the discussion. They might do a little extra homework and bring more to the conversation than you expected.

The best way to identify needs is to drill and run evolutions. As the officer you will get a really good idea of where you need to improve. We'll discuss it later, but you need to have some general benchmarks in place to really make this effective and the drills need to focus on one or two areas at a time.

When you run these drills and evolutions, the members will recognize where their struggles are and often will offer ideas and solutions to improve. They challenge themselves to do better when they understand what you are trying to accomplish. Let them use their own perspectives to help make the needed changes and allow them to take some ownership of the team.

Don't forget to be a part of the process, not just an observer. If the crew needs to improve, that includes you. Too many times we see officers not including themselves in drills and evolutions so they can monitor and evaluate the performance of the crew. Although this does need to be done, you can't completely fix any problems without throwing yourself into the mix.

More than likely your department is like most and your manning and staffing can change from day to day or hour to hour if you're a volunteer. Most of us have a minimum crew size and a maximum crew size when considering our first-due response. It's easy to get locked into the "normal" crew size and drill just with those numbers. To truly be prepared and proficient, we have to drill with all of the variables. You will learn fast that as challenging as having too few members can be, having too many can be just as challenging if we don't train to utilize those members in the most efficient ways possible. It is critical that we drill for all of the variables we could be working with.

Using multiple variables of personnel also affords you the opportunity to do the same drill in many different configurations. Needing to drill with different numbers will let you make easy variances to the same drill and to focus on the most deficient area. You can figure out specifically what number is the most challenging. Additionally, personnel may be called to perform different tasks depending on how many firefighters you're working with at that given time, requiring them to be proficient in more than one role.

By identifying the needs of your company, you will be better able to create and deliver quality, meaningful, and realistic drills for your members. Couple that with being a part of the process and using multiple variables, and your members will enjoy company drills and feel more prepared due to your diligence.

Drill with Purpose

After identifying needs, the company officer has to ensure that drills and training evolutions have purpose. We've all been there: you go out to drill and it is so mundane and simple and it has been done so many times that everybody just goes through the motions. There is no meaning or purpose for the drill or training; it is being done because that is what is supposed to happen.

The first part of drilling with purpose is to set obtainable and operationally pertinent benchmarks. There really isn't a perfect formula but the desired outcome must involve more than just completing the task successfully. At some point you have to look at what you are trying to accomplish and what parameters you want to operate within.

Using time is a great benchmark for company drills. It creates a little pressure and it is easy to measure. Additionally, it provides a real, tangible result for your crewmembers to look at and strive to improve on. You definitely want to pull the line, force the door, or perform any other task or skill correctly, but you have to look at the efficiency involved as well.

Your benchmarks should be based on your standard operational guidelines, your available resources, and standards and best practices. Some are going to be unique to your company and department, based on a lot of unique factors that you will have to learn and study for. Generally speaking, the best benchmark for successful drilling is time: How long does it take us to correctly deploy a 200-foot preconnected line to the front door of a building with a 50-foot setback? (That should include masked up and flowing water.) If you don't make your drills and benchmarks that specific, the drills will slowly drift into generalities.

In addition to the time and specifics, you must ensure that you are meeting the expectations and requirements of your standard operational guidelines (SOGs). If your department has adopted them, you have to follow them, whether you like them or not. If you don't like them, this is a perfect opportunity to change them! Go out and drill on the SOGs you aren't sold on. Be specific with your drills and evolutions with very specific benchmarks that fall in line with the SOG. Be comprehensive and thorough with this and you can test your way against the SOG. You may find that the SOG isn't that bad or you may find that it is lacking. Since you were very specific with your drills, you will be able to easily explain the deficiencies in the SOG and where it needs to be improved. (By the way, you can also offer your alternate drills and those results to back up your findings.)

By including specific, predetermined benchmarks and having a solid understanding of your SOGs, you can create more meaningful drills for your companies. When it's required, you can also be a positive change agent who is the standard for other officers to follow.

Be Patient

The new company officer is eager and driven. They want to get things done and they want to get them done now! This can create unrealistic expectations for their members to try and meet, increasing or creating some tension between the new officer and his crew.

In training and drills, the company officer must resist the urge to get everything done as quickly as possible. There has to be some patience when identifying the needs and evaluating the skill levels of members. Pushing too hard too soon will create problems with the members and you will have to work much harder than you should have to.

Something to keep in mind is that crewmembers don't know what they don't know. Don't assume that they are going to immediately adapt and meet all of your expectations regarding their skills. This is why the drills and evolutions for determining needs is so critical. It's not fair to expect results and outcomes from your new crew without allowing them some time to learn and adapt.

Be patient with progress and monitor how fast you move. Don't overload the schedule and burn your crew out by doing too much, too fast. A good pace is critical and also allows time for improvement. If we don't have some patience, our members never really get the time to make the adjustments to improve and make the needed changes.

The Basics

There many "sexy" topics in the fire service that get a lot of attention when it comes to training, and most of them have absolutely nothing to do with fighting fire. I don't have anything against technical rescue (I'm a rescue tech), hazardous materials, water rescue, or any other ancillary service that some departments offer, but those aren't what we do day in and day out.

When we are looking at creating company drills, we have to put a huge focus and emphasis on the basics of what we do: firefighting. Fighting fires is what our citizens expect us to be good at. No, they expect us to be *great* at fighting fires. They want their property saved and they definitely expect a great deal of expertise at saving *them* if they are trapped. Being great at fighting fire means we must spend a large amount of our time concentrating on the basics.

The basics are the tasks and skills required to do the job we signed on to do: Packing and pulling hose. Forcing doors with hand tools. Carrying and raising ladders. Pumping our engines and securing a water supply. Correct orientation and search of a room. Knowing our flows and stretch lengths. Preplanning for fire in our response area. These are just a few of the functions we must do often to be great at fighting fires.

We cannot perform these skills once in a while and expect to be great at them. So yes, we will be doing these over and over again, but we can add twists here and there to keep it from getting mundane and monotonous. Placing the emphasis on the basics in your drills is the fastest way to be a great company and to perform at a high level on the fireground.

Don't ignore the other jobs you perform as a fire department; you must stay sharp in everything you do. But the basics must come first. If you forget what your primary mission is and drift too far to one direction or the other, you will not put the time and effort needed into the skills you are most likely to need and utilize when the bell hits.

Make the basics the foundation of your drills and practice them frequently, and you can expect great results on the fireground. You will make your citizens proud, and that's what really matters.

Be Opportunistic

So many company officers complain about not having time or having too much on the calendar for drilling. Although this can seem true and some may have administrations that put the focus on things other than preparing for doing the job the public expects, there are many ways to drill when it's not scheduled or your time is tight.

Be creative every time you're on the road. No matter the reason for being out and about, except for running emergency calls, look for opportunities to drill. As a captain I looked for things to do every time we were out. It doesn't take a lot of time and it presents great opportunities for a nonsterile environment with live, realistic conditions.

One example would be responding to an automatic fire alarm, especially when it's after hours. Although many have key boxes for access, sometimes you have to wait for a key holder. After ensuring there is not an immediate threat identified, use the time to drill. Look at the doors and talk about how you would force them. You can train your apparatus driver to spot buildings in the dark while you check the roof on a call. It's a great time to practice those types of skills. Stretch a line to multiple doors to practice determining line selection. As you can see, you're only limited by imagination and effort. Normally it's the effort that some are lacking.

We did this on a reported kitchen fire that we were second-due on. The first-due company reported the fire was out with minimal damage and all other units could return. I told my operator to pull to the side of the building and lower the ladder rack. The building had a parapet wall and it had been a while since I had laddered one without an aerial device and my firefighter had never done it, so this was perfect. I communicated to the first company that we were checking the roof and we set up in some bushes, creating an obstacle for us. It was a great discussion opportunity. We checked the roof and learned something in the process. It took all of 10–15 minutes from start to finish. But it was a great experience for all of us.

You can easily stretch rope or lines to determine hose stretch lengths at apartment buildings. You can pull behind strip malls and commercial buildings and check hydrants and spot fire department connections (FDCs). Drive through neighborhoods estimating hydrant distances and discuss water supply alternatives.

Don't limit your drills to just when they are scheduled or on the calendar, or you'll be missing out on some great drills and crew discussion.

11

10 COMMANDMENTS FOR COMPANY OFFICERS

Through my years in the fire service I have served in just about every capacity, between my career departments and my volunteer agency. There is no doubt in my mind that the company officer can have the biggest impact on the department and its successes. This section is a summary of the book and specifically the chapter on leadership. I felt it was important to break down some of the finer and most important parts of the book into my 10 Commandments for the Company Officer.

The company officer is involved in so many functions and decisions daily that it puts them in a place to have an enormous impact on their crews and the department as whole. From mentoring, drilling, running, and operating at fires and emergency scenes to helping create SOGs, the company officers are on the front lines getting it done (fig. 11–1).

On the flip side, however, they can also be the biggest obstacles for any fire department to progress and move forward. Company officers who don't take their positions seriously or those who stop working after the promotion can be their departments' biggest problems.

My experiences, along with great mentors and advisors, have taught me some valuable lessons; I learned what worked and what didn't as a company officer. Having experience in both volunteer and career departments, these 10 commandments will work in any type of system. There are no magic bullets and you won't be perfect in every situation. But I learned that if I followed these 10 simple items on a daily basis, things typically went well and I was aware of what I was doing and how I was doing it.

Fig. 11–1. The company officers are on the front lines, getting it done. (Courtesy Engineer James Bell, Metro West Fire Protection District.)

These 10 commandments are suggestions and not necessarily what will work for every officer at every department. (We'll discuss that more in our list!) But, when addressed and used routinely, they can make you an effective and productive company officer who garners the respect of your peers and superiors alike.

1. The Fire Officer Shall Be Prepared

I mentioned earlier in the book how to prepare to be an attractive candidate for promotion. Much of what I listed in that section applies here, but it should be mentioned again. I just can't emphasize enough the importance of preparation at all levels of the fire service.

I commented that not every one of these commandments might work for you personally or for your department, although for the most part I believe these are effective for any situation and any department. But what I meant by that is that officers need to be critically aware of where they work and for whom they work. You must know your department and what its mission is.

In your daily activities you must prepare for your crew's success based on what the department needs and wants. You must understand and know the priorities and boundaries that have been set by your department. It's not always what you want it to be and it's not always easy, but your job is to represent your department and carry out orders.

If you prepare your crews for this, it helps to create consistency and allows you to better prioritize your shifts. When we do work for those departments we don't agree with in regard to those priorities and orders, we are able to more effectively institute change when we can show that what is being done is not working. Follow orders, document the challenges, find a better way, and submit a proposal based on fact rather than opinion and accompanied by a solution.

When preparing for your day and your crew, you must create and maintain productive habits. This must be daily and consistent. This will set the tone for your crew and will be the persona you take on as an officer. Being productive does not mean doing work for the sake of doing work (what is termed "busy work"). Being productive from the company officer perspective is getting what needs to be done, done. It is based on priorities and needs, not wants. A productive company puts the priorities in place and doesn't make excuses for doing work that other crews will not or cannot get done. Being productive considers the goals and mission of the department and that of the crew.

I always looked at it from this perspective: If we don't do it, who will? It's easy to pass work along to others or to put it off. The productive company officer not only requires that necessary work and tasks be completed, they also instill a sense of pride in the crew that "we get done what others don't."

This leads us into the next section of setting expectations. Being productive is critical for success, and the best way to ensure that you and your crew stay productive is to set expectations.

2. The Fire Officer Shall Set Expectations—And Follow Through

I have already dedicated an entire chapter to this, but it goes into the 10 commandments and should be mentioned again. This was a critical step for me to take and follow through on that has paid huge dividends. Without specific, well-defined, and timely expectations, you will flounder and your members will be constantly guessing what will be next and what you will require.

This is where you set your priorities for daily activities and being productive. My four priorities, in this order, are the citizens, the department, the crew, and then the individual. You will find that establishing these priorities helps you and your crew manage the daily activities and creates a mindset that is lived every shift. But you must require and make decisions based on these priorities for them to be effective. That means that in some or many cases you may have to tell crew members no and make them perform tasks that have in the past been unpopular.

As was mentioned in chapter 3, be productive, be generous, be constructive, be professional, and be aggressive!

3. The Fire Officer Shall Know Their Crew

As the officer you are now responsible for more than just yourself. You must learn who your members are. This is the only way to truly create trust and build those relationships that are so valuable to our profession. Additionally, we need to know what motivates and moves our members so that we can help them succeed.

We use this in many aspects of leading and managing our members and we have to, at times, get a little uncomfortable in order to make this work. Learn what their interests are, the names of their family members, and what they want from the job. The better you know your people, the better you can lead, mentor, advise, train, and learn from them. Don't discount the importance of relationships when considering your daily interactions. In all of this, your regard for them must be genuine and real. It has to be attention to small details and conversations, not just passing, cliché remarks.

4. The Fire Officer Shall Serve at All Times

I like to say that the higher up you go in rank, the more members you are obligated to serve. As the company officer you are obligated to serve the public, your superior officers, the department, and especially your crew members.

You should be looking for ways to serve your members every opportunity you get. Putting their needs ahead of your own creates trust and builds closer relationships. It also lets you truly mentor and lead at every level. This is where you really mold and shape future leaders and officers. You will create an environment of service that will spill over into everything your crew does, which benefits the crew, the department, and, most importantly, the public.

5. The Fire Officer Shall Listen

As the officer and leader, to be effective consistently, you must be a great listener. Communication starts and ends with listening to the message. The company officer must be able to offer members their full attention and consideration when they have questions, concerns, and suggestions.

Listening actively requires that you provide your full attention to your members to hear their needs. Actively listening means you are engaged in the discussion as well as paying attention to the body language and facial expressions of the person you're communicating with. Sometimes how somebody says something is more important than what they say. The only way to figure that out is to pay attention to them, not your phone or computer. When you are actively listening you are less likely to jump to conclusions and to respond inappropriately. Be patient and respond appropriately. Sometimes that may mean that your response will take time. This should be communicated to the member along with an expected response.

Finally, communicate face-to-face as much as possible and be a fact finder. Don't overreact and don't believe everything you hear at the kitchen table. Take the time to get the real and whole story.

6. The Fire Officer Shall Display Humility

This is another topic that was addressed in more detail in chapter 8. But, like other traits and commandments, this one is critical for your success as an officer and leader. What needs to be considered most is that being humble is something that builds the trust of your crewmembers. In fact, it may be one of the quickest ways to start earning that trust.

Being humble is obviously important when it comes to taking responsibility for your mistakes. But I believe one of the more critical factors of being humble is its role in allowing you to improve. Basically, if you can't admit you are deficient in areas or have made a mistake or are struggling at something, you will never address it. Being humble allows you to correct mistakes and to improve as an officer and leader. It also shows your members that you are not above the rules that apply to everyone else and that you are transparent with your own struggles. It's also a great way to share your lessons with others as you are going through them.

7. The Fire Officer Shall Demonstrate Leadership

I'm going to list a few items here for emphasis on leadership for the company officer and leaders. This is topic is extensively covered in the leadership chapter, but is definitely worthy of a spot in the 10 commandments and deserves mention. These are self-explanatory but critical reminders for you to write down and address every day:

- Always be honest, even when it's most difficult and with people you've been friends with for a long time.
- Build trust at every opportunity through relationships.
- Keep your emotions in check—very few decisions made when you're mad or euphoric end well. It's normal to have strong emotions, you just shouldn't make acting on them normal practice.
- Be the boss! You must be in charge when it's necessary. Don't waffle and don't delegate things that carry a great deal of risk to avoid consequences. Many times being the boss also means saying no.
- Don't use a one-size-fits-all approach. You're dealing with people who are motivated and learn differently. This relates back to knowing your people.

- Be able to offer and accept critical advice that is constructive. There is no reason to demean or embarrass your members when they mess up or struggle.

8. The Fire Officer Shall Make No Excuses

I'll keep this short and to the point: if you make excuses, you are a weak leader. Yes, you are a weak leader and you will struggle mightily in your position. Your members will not trust you, and neither will your superior officers. Soon you will see the same from your members and this is where a culture is created. Unfortunately, it will be a culture of excuses, lacking accountability and responsibility. Once you build it, it can take a long time to tear it down and rebuild.

Simply, if it is broken, fix it. If there are issues, address them head on. If you did it, own it and find a solution. If your members offer excuses, don't accept them. We have to be accountable to our members and ourselves. The people we serve expect us to have integrity and those we work with expect us to be responsible.

9. The Fire Officer Shall Be the Best Right Now

We discussed this in the opening chapter, but please add it to your list. I see so many future and current officers lose their perspective and violate this commandment, and some don't really know they're doing it. Many are in such a hurry to promote and move on to the next level or rank they forget about their current position and job. They are working so hard at being the best at the next rank that they neglect the duties of their current role. This is especially true for company officers who are working toward a staff chief, division chief, or battalion chief rank. They get so wrapped up in attempting to do everything they can to be promoted that they stop being the best at what they are now. I can honestly say that as a person who has sat in on promotion processes, I like to see the best at what they do come through and get promoted. I want a firefighter who is the best firefighter; they work at it and are proud to be in that position. Likewise, I want a captain who concentrates on being the best captain possible, not on being the next best battalion chief.

When you are not working at being the best at your current position, you are selling yourself short and letting your members and peers down. They will have to pick up the slack created by your inattention to your current position.

10. The Fire Officer Shall Do What Others Don't or Won't

Finally, do the little things. Here is a short list of examples of things you can do regularly to set yourself apart as a great leader and officer. Some of these were listed in other chapters, but they are critical components of our 10 commandments.

- Solve problems and chase trouble; if you walk past a problem you are condoning it and making it acceptable.
- Don't take the easy way out. When it's late, cold, hot, or uncomfortable, it's not the time to be complacent or to slack off. Push through adversity and get it done.
- Find solutions to problems. Don't be a complainer, be a fixer.
- Define your legacy with the little things you do daily. What will they say when you're gone? What is your contribution?
- Do the right things, the right way, for the right reasons, all the time!
- What you do when nobody is looking or can't see is who you are.

So, there it is and I hope you find the list helpful in summarizing many components of the book. You can probably come up with your own ideas and items to add or replace on this list and that's great! The key is to be methodical in how you operate and to lead with purpose. Use these action items to mentor future officers and to keep your influence in the right direction.

NOTES

NOTES

NOTES

NOTES